# NEW GEOGRAPHIES

## 09
## POSTHUMAN

Edited by
Mariano Gomez-Luque
& Ghazal Jafari

09
POSTHUMAN

New Geographies 09
**Posthuman**

**Editors**
Mariano Gomez-Luque & Ghazal Jafari

**Editorial Board**
Daniel Daou, Daniel Ibañez, Ali Fard, Nikos Katsikis, Taraneh Meshkani, Pablo Pérez Ramos

**Founding Editors**
Gareth Doherty, Rania Ghosn, El Hadi Jazairy, Antonio Petrov, Stephen Ramos, Neyran Turan

**Advisory Board**
Eve Blau, Neil Brenner, Sonja Dümpelmann, Mohsen Mostafavi, Antoine Picon, Jennifer Sigler, Charles Waldheim

**Academic Advisor**
Neil Brenner

**Editorial Advisor**
Jennifer Sigler

**Production Advisor**
Meghan Sandberg

**Text Editor**
Kari Rittenbach

**Proofreader**
Elizabeth Kugler

**Graphic Designer**
Sean Yendrys

**Design Production Assistance**
Bo-Won Keum

**Cover Illustration**
Krystelle Denis

© 2017 by the President and Fellows of Harvard College and Actar D. All rights reserved. No part may be reproduced without permission. Text and Images © 2017 by their authors.

Printed in Estonia by Printon.

Logo design by Jean Wilcox.

ISBN: 978-1-945150722
LCCN: 2017955395

www.gsd.harvard.edu/newgeographies

*New Geographies* is a journal of Design, Agency, and Territory founded, edited, and produced by doctoral candidates at the Harvard University Graduate School of Design. *New Geographies* presents the geographic as a design paradigm that links physical, representational, and political attributes of space and articulates a synthetic scalar practice. Through critical essays and projects, the journal seeks to position design's agency amid concerns about infrastructure, technology, ecology, and globalization.

*New Geographies 09: Posthuman* has been made possible with support from the Graham Foundation for Advanced Studies in the Fine Arts and the Harvard GSD Office of the Dean.

All attempts have been made to trace and acknowledge the sources of images. Regarding any omissions or errors, please contact:

New Geographies Lab
Harvard University Graduate School of Design
48 Quincy Street
Cambridge, MA 02138

**Distribution**
Actar D
440 Park Avenue South
New York, NY 10016
actar-d.com

Special thanks to: Chief Lady Bird, Aura, fifth and sixth grade students at First Nations School, Benjamin Grant, Cassio Vasconcellos, Krystelle Denis, Michael Hansmeyer, David Maisel, Robert Burley, Aurélien Maréchal, and Extramural Activity. In addition, we would like to thank the Harvard GSD Dean's Office, the Harvard GSD Publications Office, text editor Kari Rittenbach, graphic designer Sean Yendrys, and all the people who have helped us in the process of making this journal, especially our thesis advisors Neil Brenner and Pierre Bélanger.

| | | | |
|---|---|---|---|
| 8 | Mariano Gomez-Luque & Ghazal Jafari<br>**Posthuman** | 140 | Eli Nelson<br>**"Walking to the Future in the Steps of Our Ancestors": Haudenosaunee Traditional Ecological Knowledge and Queer Time in the Climate Change Era** |
| 12 | Rosalind Williams<br>**Redesigning Design** | 150 | Charles Waldheim<br>**Wild Life** |
| 20 | Erik Swyngedouw<br>**More-than-Human Constellations as Immuno-Biopolitical Fantasy in the Urbicene** | 156 | John Dean Davis<br>**The Cyborg in the Garden** |
| 28 | Benjamin H. Bratton<br>**Geographies of Sensitive Matter: On Artificial Intelligence at Urban Scale** | 164 | Namik Mackic & Pedro Aparicio Llorente<br>**Horizon of a Different Machine: Geotechnicity** |
| 34 | Luciana Parisi<br>**Against Nature: The Technological Consciousness of Architectural Design** | 177 | An Interview with Cary Wolfe<br>**Critical Ecologies of Posthumanism** |
| 45 | Barbara Adam<br>**Time by Design** | 186 | McKenzie Wark<br>**Adventures in Third Nature** |
| 48 | The GIDEST Collective<br>**Living Past the End Times** | 194 | Jason W. Moore<br>**Confronting the Popular Anthropocene: Toward an Ecology of Hope** |
| 54 | Shannon Mattern<br>**Extract and Preserve: Underground Repositories for a Posthuman Future?** | 203 | **Visual Essay Citations and Image Credits** |
| 64 | Antoine Picon & Carlo Ratti<br>**Mapping the Future of Cities: Cartography, Urban Experience, and Subjectivity** | 204 | **Biographies** |
| 70 | Alejandro Zaera-Polo<br>**The Posthuman City: Urban Questions for the Near Future** | | |
| 81 | A Conversation with Eyal Weizman<br>**Beyond the Threshold of the Human** | | |
| 90 | Stephen Graham<br>**Satellite: Enigmatic Presence** | | |
| 98 | Martín Arboleda<br>**On the Alienated Violence of Money: Finance Capital, Value, and the Making of Monstrous Territories** | | |
| 108 | Mimi Sheller & Esther Figueroa<br>**Geopolitical Ecologies of Acceleration: The Human after Metal** | | |
| 117 | Jose Ahedo<br>**Animal Life: A Visual Essay** | | |
| 132 | Rosetta Elkin<br>**Plant Life: The Practice of "Working Together"** | | |

# Posthuman

# Mariano Gomez-Luque & Ghazal Jafari

I

At the turn of the second millennium, media theorist Katherine Hayles argued that the "historically specific construction" we know as the "human" was then giving way to another, called "the posthuman." She located the initial coordinates of this transition in mid-20th-century cybernetics, a "breathtaking enterprise . . . nothing less than a new way of looking at human beings." For Hayles, the emergent posthuman subject differed radically from the liberal subject that had been the presumptive model of the human ever since the Enlightenment.[1]

The unremarked transition from human to posthuman nevertheless prefigured the foundational critique of different strands of humanism in the late 1960s, leading to a variety of new fields of study that challenged humanist interpretations of human existence.[2] These developments cumulatively triggered what many scholars have since described as a "posthuman turn"—characterized by the concern for, first of all, upsetting the normative conventions that position Western "man" as the universal bearer of the human; and second, countering the hegemony of *anthropos* relative to other forms of (nonhuman) life. "Posthumanism" thus appeared as a new critical epistemology that not only combined a variety of anti-humanist and post-anthropocentric positions, but also attempted to exceed the terms of this binary scheme.[3]

Today, the process of decentering of the human unleashed by this posthuman turn is further exacerbated by an omnipresent sense of crisis that has transpired in close conjunction with a series of radical scientific, technological, and spatial transformations. Indeed, a cascade of intensifying crises—environmental, ecological, geopolitical, economic, humanitarian—together with increasingly sophisticated socioeconomic modalities of violence, brutally imposed on larger and larger segments of the global population, pose unprecedented challenges to human life on the planet.[4] In parallel, scientific advancement and rapid technological change are modifying the very parameters through which long-standing definitions of the human were constructed, as much as ubiquitous urbanization is altering the environments in which social life historically unfolded.[5] These circumstances define the contours of a posthuman condition; a historical formation which, far from being the *n*th variation in a long sequence of prefixes, instead underlines the urgency of critically rethinking the ways of *being* in the world that are currently emerging.[6]

Despite perplexing anxiety regarding the place of the posthuman subject within a rapidly changing global context, it is important to remember that the human—as Foucault in particular argued—was never a neutral or universal category.[7] Rather, it is a historically constructed concept that indexes access to power, entitlement, and privilege. Certainly, as philosopher Rosi Braidotti affirms, the same applies to the category of the posthuman.[8]

Put differently, the posthuman condition does not take place in a vacuum, but crystallizes within the political economy, and in relation to the post-anthropocentric technologies of, contemporary biogenetic capitalism.[9] Armed with a robust technoscientific apparatus spanning the core fields of biotechnology, nanotechnology, information technology, and cognitive neuroscience, biogenetic capitalism invests in the control of the informational power contained in the genetic code of all living matter: human, animal, bacterial, even the mineral world. Biogenetic capitalism thus reduces both human and nonhuman life to mere material for technoscientific manipulation, potentially subjecting it to hitherto unthinkable forms of control, domination, and instrumentalization.[10]

In this context, it is clear that to be posthuman—that is, to be a subject of our time—does not necessarily imply that one is "post-power, post-class, post-gender, post-imperial, or post-violence."[11] Quite the opposite: the posthuman signals a type of subjectivity deeply embedded in the neoliberal governance and corporate-managerial practices of the contemporary world order.

While addressing the complex dimensions underpinning the current historical milieu, posthuman thought is nevertheless driven by an ethico-political project. "Becoming posthuman," Braidotti argues, involves the possibility to not only "decide together what and who we are capable of becoming" but also "for humanity to re-invent itself affirmatively, through creativity and empowering ethical relations, and not only negatively, through vulnerability and fear." What is more, embracing the posthuman condition and its historical and theoretical dimensions offers "a chance to identify opportunities for resistance and empowerment on a planetary scale."[12]

At the same time, as theorist Cary Wolfe suggests, posthuman*ism*, as a philosophical framework, goes beyond the chronological succession implied by the prefix "post-." Posthumanism is not only concerned with the present historical subject, in the present historical situation, but more fundamentally, with "what thought has to become" in order to confront the daunting challenges of our era.[13]

II

This ninth issue of the journal *New Geographies*, titled *Posthuman*, surveys the urban environments shaping the more-than-human geographies of the early 21st century. Seeing design as a geographical agent deeply involved in the territorial engravings of contemporary urbanization, *New Geographies 09* embraces the "planetary" as the ultimate spatiotemporal stage of the posthuman condition.[14]

This interpretation is fueled by awareness of the historical instrumentality of both geography and design (as disciplinary fields and spatial worldviews) in the delineation and pursuit of new "frontiers" serving the ambition for endless expansion of the *human empire*.[15] That is, geographic knowledge and design strategies, methods, and metrics applied in the organization of global space have been crucial for the "territorial acquisition, economic exploitation, militarism, and . . . practice of class and race domination"[16] which characterize imperialist power.[17]

With this in mind, geographic and design thinking are here mobilized in a different direction: namely, as an interpretive lens through which to trace how those crises and historical circumstances that have destabilized the inherited schema of the human manifest themselves spatially—how they are indexed by the complex geographical formations of the contemporary built environment.[18]

This issue of *New Geographies* gathers together contributions that critically evaluate a wide array of manufactured territories that now cover the surface of the planet, and extend even further beyond its geophysical boundaries into outer space. These include the technological environments, operational landscapes, underground facilities, outlying airfields, infrastructural networks, and other "third natures" which, embodying the manifold entanglements between humans and nonhumans, define the hybrid geographies of the urban world.

Far from exhaustive or conclusive, the present volume has as its main objective to propose a more accurate depiction of the intricate cultural, biopolitical, economic, and territorial grounds from which a genuinely posthuman spatial condition is materializing. The spectrum of positions assembled in *Posthuman* reflect—sometimes from divergent standpoints—the fervent debate surrounding the conceptual, epistemological, and historical dimensions of such emergent spatial configuration.

### III

The diverse contributions to this volume are loosely organized, constellating a mosaic of critique, speculation, dialogue, and narrative. Complementing this structure, **Krystelle Denis**'s cover design conveys the depth of those intertwined organic and machinic ecologies that constitute today's planetary geographies.

The first set of textual contributions explores two key themes: the role of design in the contemporary world, and the problem of "design intelligence" in the age of artificial intelligence (AI).[19] **Rosalind Williams** begins the discussion with a critical assessment of the role of (and frontiers opened by) design in the profound spatial transformations across the built environment during the last two centuries. **Erik Swyngedouw** mobilizes the concept of the "Urbicene"—a coinage that implicates the primary site of impact in the Anthropocene—in relation to more-than-human ontologies, and considers their ramifications for current discourses on the urban. **Benjamin Bratton** speculates on the implications of AI at a geographical scale, while **Luciana Parisi** contends that through advanced computation techniques, architectural design engenders its own mode of (inhuman) thought, or "technological consciousness."

As speculative passages between the first and second sections of the journal, **Barbara Adam** reflects on the changing rhythms of contemporary life inflected by the increasing commodification and colonization of time, while the **GIDEST Collective** envisions a fictional scenario in which an alternate past is recovered in a near future and plied to the disorienting coordinates of the present.

The second series of articles traces the influence of contemporary urban systems, within which new spaces, subjectivities, technological agents, and cultural identities take form. **Shannon Mattern** investigates the shifting usage of the underground in our "age of anthropogenic geoengineering and posthuman intelligence"; **Antoine Picon and Carlo Ratti** review emergent forms of subjectivity engendered in the novel technologies of urban cartographic systems; and **Alejandro Zaera-Polo** reconceptualizes a 21st-century urban cosmology in the "Posthuman City."

A conversation with **Eyal Weizman** concentrating on his recent work in Forensic Architecture sets the terms of another important topic linked to the posthuman condition: the many forms that the *inhuman(e)* adopts in the context of global and technologically mediated societies.[20] **Stephen Graham** writes on the political geography of inner and outer space, using the satellite as a case study to vividly portray a new sense of "vertical free fall"; **Martín Arboleda** links financial extraction and its "monstrous" territories to the schism between money's *bad infinity* and the embodied realities of human and ecological existence; and **Mimi Sheller** describes the environmental risks posed by large-scale aluminum industries in the drive toward planetary urbanization.

A visual essay on the "new domestic frontier" of animal environments, by **Jose Ahedo**, precedes the texts of the third section, which engage with the problematic of the *other*.[21] **Eli Nelson**, focusing on the recent history of the Haudenosaunee Confederacy, describes how Western culture misconstrues or dismisses the practices of "racialized others," such as indigenous traditional ecological knowledge (TEK), which models alternate concepts of sovereignty and futurity; **Rosetta Elkin** traces the way in which the "naturalized other"—that is, plant life—has been exploited as both o*bject* (fixed, static, non-sentient) and *objectified* knowledge; **Charles Waldheim** tours the technomanagerial systems that mediate occasional encounters between humans and wildlife within the engineered space of the contemporary airport; **John Davis** gives a critical reading of the cyborg, or "technological other," as it has been (mis)appropriated in the landscape design imaginary; and **Namic Mackic and Pedro Aparicio Llorente** propose that the earth itself is a geologically and materially complex "perpetual machine" that both produces and supports life by means of geotechnicity.

The journal's final interlude features a conversation with philosopher **Cary Wolfe**, in which various theoretical dimensions of posthumanism are taken up in connection with the social, political, and ecological challenges facing us today.

Finally, the closing essay cluster addresses another foundational proposition of the posthumanist approach: the nature-culture continuum, seen against the backdrop of the relentlessly ongoing (planetary-scale) production of nature.[22] **McKenzie Wark** ruminates on "third nature," the vast, abstract computational veil that wraps the planet in information and defines the material and infrastructural strata that today organize human and nonhuman life; and **Jason Moore** challenges the Cartesian divide still conceptually ingrained in the "popular Anthropocene," instead proposing "new ecologies of hope" through which to think about *nature* in terms of "an ethics of care, for humans, for the web of life, and for the multispecies interdependencies that make the good life possible."

IV

Drawing on a variety of scholarly expertise, from the fields of architecture to urban theory, from landscape to ecological thought, from philosophy to infrastructure to media studies, *New Geographies 09—Posthuman* stimulates wide-ranging debate on the potential for design to engage with the complex spatiality, more-than-human ecology, and diverse life-forms that define a different kind of planetary environment: one in which the human, per Cary Wolfe, finally acknowledges never having been "master in its own house." Consequently, *Posthuman* advocates that the challenge to build a more socially and ecologically just urban world—inhabited by human and nonhuman subjects—be brought to the surface of the political imagination. In this regard, the posthuman turn must be considered an open-ended project, one suitable for considering not only "what and who we are capable of becoming," but also what kind of worlds we dare to envision, and may collectively create.

---

1   N. Katherine Hayles, *How We Became Posthuman: Virtual Bodies in Cybernetics, Literature, and Informatics* (Chicago: University of Chicago Press, 1999).

2   Critiques of humanism stemmed from new academic disciplines including, among others, gender studies; feminism; cultural studies; postcolonial studies; ethnicity, race, and migration studies; media and new media studies; and human rights studies. See Rosi Braidotti, *The Posthuman* (Oxford: Wiley, 2013), 13–54.

3   See Braidotti, *The Posthuman*, 13–54 and 143–85; and Cary Wolfe, *What Is Posthumanism?* (Minneapolis: University of Minnesota Press, 2009), xi–xxxiv.

4   See Jonathan Crary, *24/7: Late Capitalism and the Ends of Sleep* (London: Verso, 2013). On new forms of brutality, see Saskia Sassen, *Expulsions: Brutality and Complexity in the Global Economy* (Cambridge, MA: Harvard University Press, 2014).

5   Some examples include: recent progress in the fields of artificial intelligence (AI) and brain-computer interface (BCI) technology that have caused experts to suggest an "intelligence explosion" may occur relatively soon; the profound impact that digital technologies and increasing automation have on almost every aspect of contemporary social life; and the global socio-spatial transformations introduced by intensifying urbanization. See, respectively: Nick Bostrom, *Superintelligence: Paths, Dangers, Strategies* (Oxford: Oxford University Press, 2016); Adam Greenfield, *Radical Technologies: The Design of Everyday Life* (London: Verso, 2017); Neil Brenner, ed., *Implosions/Explosions: Towards a Study of Planetary Urbanization* (Berlin: Jovis, 2014).

6   "[T]he posthuman condition introduces a qualitative shift in our thinking about what exactly is the basic unit of common reference for our species, our polity, and our relationship to the other inhabitants of this planet." Braidotti, *The Posthuman*, 1.

7   See Michel Foucault, *The Order of Things: an Archaeology of the Human Sciences* (London and New York: Routledge, 1989).

8   Rosi Braidotti, Keynote Lecture, Posthumanism and Society Conference, New York University, New York, May 9, 2015, www.youtube.com/watch?v=3S3CuINbQ1M.

9   See Braidotti, *The Posthuman*, 55–104. On contemporary capitalism, Braidotti draws on Moulier-Boutang's notion of "cognitive capitalism." See Yann Moulier-Boutang, *Cognitive Capitalism*, trans. Ed Emery (Cambridge: Polity Press, 2012).

10   Braidotti, *The Posthuman*, 55–104. [Ch. 2]

11   Braidotti, Keynote Lecture, Posthumanism and Society Conference, New York University, New York, May 9, 2015, www.youtube.com/watch?v=3S3CuINbQ1M.

12   Braidotti, *The Posthuman*, 195.

13   Cary Wolfe, *What Is Posthumanism?*, xvi. While Hayles and Braidotti each provide a historical and theoretical account of the emergent posthuman subject, Wolfe instead emphasizes the posthumanist mode of thought, hence the title of his book.

14   "Planetary," as mobilized here, is not meant to imply a 'fix scale' but rather a signifier for the 'scaleless,' or the 'trans-scalar.' See Ross Exo Adams, "On Scaleless Urbanization: Cybernetic Infrastructures, Resilient Design and the Becoming of Planetary Space," in *Infrastructure Space*, eds. Ilka & Andreas Ruby (Berlin: Ruby Press, 2016), 229–37.

15   See Rosalind Williams's essay in the present volume, "Redesigning Design," 10–15.

16   Brian Hudson, "The New Geography and the New Imperialism." *Antipode* 9, no. 2 (1977):12.

17   On the role of geography and design in imperialist and colonialist practices, see Felix Driver, "Geography's Empire: Histories of Geographical Knowledge," *Environment and Planning D*, 10 (1992): 23–40; William Cronon, "The Trouble with Wilderness: Or, Getting Back to the Wrong Nature" in *Uncommon Ground: Rethinking the Human Place in Nature*, ed. William Cronon (New York: Norton, 1995): 69–90; Patrick Wolfe, "Settler Colonialism and the Elimination of the Natives," *Research Network in Genocide Studies* 8, no. 4 (2006): 387–409; W. J. T. Mitchell, *Landscape and Power* (Chicago: University of Chicago Press, 1994); Eyal Weizman, "Are They Human?" *E-Flux Architecture: Superhumanity* 51 (October 2016), www.e-flux.com/architecture/superhumanity/68645/are-they-human; Pierre Bélanger, ed., *Extraction Empire: Sourcing the Scales, Systems, and States of Canada's Global Resource Empire* (Cambridge, MA: MIT Press, forthcoming).

18   This geographical complexity has been exacerbated by an arsenal of novel spatial technologies that aim to operationalize space at all scales, from the organic composition of matter to the atmospheric realm. See Beatriz Colomina and Mark Wigley, *Are We Human?: Notes on an Archaeology of Design* (Zurich: Lars Müller Publishers, 2016).

19   Posthuman theory conceives of intelligence, "thinking," and more generally, the capacity to produce knowledge not as the exclusive, unique prerogative of humans, but as a distributed form of cognition that encompasses all living and self-organizing matter, as well as all kinds of technological networks. Thinking, thus theorized, is what "being alive feels like." Braidotti, Keynote Lecture, Posthumanism and Society Conference, New York University; see also Hayles, *How We Became Posthuman*, 50–83, 131–59, 222–46; Wolfe, *What Is Posthumanism?*, 1–142.

20   See Braidotti, *The Posthuman*, 105–42; Achile Mbembe, "Necropolitics," *Public Culture* 15, no. 1 (Winter 2003): 11–40.

21   See Wolfe, *What Is Posthumanism?*, 99–126; Braidotti, *The Posthuman*, 13–54. Braidotti gives several useful categories for the human's others, which we have adapted in this volume: the racialized other (native and indigenous peoples), the sexualized other (women), the naturalized other (animals, the environment, earth), and the technological other (machines). See Braidotti, *The Posthuman*, 27, 94, 109.

22   "Matter is not dialectically opposed to culture, nor to technological mediation, but continuous with them." Braidotti, *The Posthuman*, 35. Importantly, this approach should not be confused with that of a "flat ontology," as political theorist Jane Bennet also argues in *Vibrant Matter: A Political Ecology of Things* (Durham, NC: Duke University Press, 2010), 110–22. For a discussion of the philosophical framework of flat ontology as it differs from posthuman theory, see Ian Bogost, *Alien Phenomenology, or What It's Like to Be a Thing* (Minneapolis: University of Minnesota Press, 2012).

# Redesigning Design

**Rosalind Williams**

Writing in 1994, in his overview of 20th-century history titled *The Age of Extremes*, Eric Hobsbawm notes a proliferation of words beginning with "post":

> When people face what nothing in their past has prepared them for they grope for words to name the unknown, even when they can neither define nor understand it. Some time in the third quarter of the century we can see this process at work among the intellectuals of the West. The keyword was the small preposition "after," generally used in its latinate form "post" as a prefix to any of the numerous terms which had, for some generations, been used to mark out the mental territory of twentieth-century life.[1]

He lists examples: post-industrial, post-imperial, post-modern, and post-structuralist, among others. Adding the prefix "post" to familiar words, Hobsbawm concludes, signifies an event of consciousness: "In this way the greatest and most dramatic, rapid, and universal social transformation in human history entered the consciousness of reflective minds who lived through it."[2]

The terms "posthuman" and "posthumanism" (with or without hyphenation) were just coming into fashion then. They have since become much more common and should be added to Hobsbawm's list.[3] Yet their meaning is cloudy, because the post-combination essentially implies a void: "I can't find words for what is going on." So what is the "greatest and most dramatic, rapid, and universal social transformation in human history"? This question, and not the specific term "posthuman," will be the focus of attention here.

In the middle section of *The Age of Extremes*, Hobsbawm explains that for him the transformation is essentially the "death of the peasantry" that took place in the third quarter of the 20th century, when the Neolithic Era, characterized by the harvesting of land and sea, ceased to be the material and cultural foundation of human life.[4] Many other historians have defined the transformation as the vast expansion of productivity in the Industrial Revolution beginning in the late 1700s, extending through the Second Industrial Revolution of the late 1800s and early 1900s, and continuing in the still ongoing Information Revolution.

More recently, other historians have borrowed terms and concepts from geology to characterize the transformation as the advent of the age of the Anthropocene. This label may be considered a bold act of disciplinary border crossing; or an admission that traditional historical language is inadequate to express the significance of new historical events. On the other hand, the language of natural history is inadequate for describing human history. Its time constants are far too sweeping to fit the temporal scales of human experience. The evidence it seeks in stratigraphic and other geological markers rarely register struggles for power within the human species, much less the role of human consciousness in co-evolving with its habitat.

All these ways of looking at "the greatest and most dramatic, rapid, and universal social transformation in human history" have this in common: they describe its essential feature as the creation of a new habitat for human existence. Since the emergence of the species, nonhuman nature has been the ground of human life. The relationship between figure and ground—or better yet, figure and surround—has shifted such that we now dwell in a hybrid environment where human-generated technologies and the given world are inextricably mixed. In this new habitat, humans initiate processes—biological, meteorological, and radioactive, to take some examples—that would never have occurred on the planet without us. We are not the only agent in our habitat, but we are the primary agent. The substance of this great transformation is humankind's creation of a new habitat, which humans dominate but do not control.[5]

How do we talk, and therefore think, about this phenomenon as a historical event? Usually we do so from the outside, using as evidence things and systems we have made and continue to make and operate: plows, mills, knives, steamboats, photographs, computers, highways, spacecraft, and so much more. The re-creation of our habitat through making and using all of these material things is a process that has taken place in multiple dimensions, with many complicated feedback loops. At certain moments, however, these processes crystallize into events that make us realize what is happening over larger stretches of time and space. These are events which, in Hobsbawm's words, show that the transformation of the human habitat has "entered the consciousness of reflective minds who lived through it."

Such an event of consciousness took place in the late 19th and early 20th centuries. Within the space of a few decades, the great, centuries-long project of the modern West—the mapping of the planet, as "mapping" was defined by the mapmakers themselves—drew to a close. World atlases from that time show just a few areas of the globe (most notably the Poles and the interior of Africa) that had not yet been surveyed. The map implicitly proclaimed these "empty places" would soon be filled in. The notable features of the lands and

---

1   Eric Hobsbawm, *The Age of Extremes: A History of the World, 1914–1991* (1994; repr., New York: Vintage Books, Random House, 1996), 287.

2   Ibid., 288. Hobsbawm's passage echoes the title of one of the best-known books on this topic, Karl Polanyi's *The Great Transformation*, first published in 1944 as *The Origins of Our Time*, with *The Great Transformation* as subtitle. The phrases were later reversed in the 2001 edition, *The Great Transformation: The Political and Economic Origins of Our Time*. This term and debates surrounding it continue in lively fashion: see Johan Schot, "Confronting the Second Deep Transition through the Historical Imagination," *Technology and Culture* 57, no. 2 (April 2016): 445–56.

3   For a summary of the intellectual history of these terms, see Cary Wolfe, *What Is Posthumanism?* (Minneapolis: University of Minnesota Press, 2010): xi–xxxiv.

4   Hobsbawm, *The Age of Extremes*, 289.

5   See discussion in Rosalind Williams, "The New Human Habitat," in *Retooling: A Historian Confronts Technological Change* (Cambridge, MA, and London: MIT Press, 2002): 19–26. Like Hobsbawm, I consider the Neolithic Age the closest reference point for the transformation of recent times (23–24). The distinction between domination and control has been explored by Michel Serres ("Our very mastery seems to escape our mastery"), especially in Serres, with Bruno Latour, *Conversations on Science, Culture, and Time*, trans. R. Lapidus (Ann Arbor: University of Michigan Press, 1995): 171–72.

waters of the globe would be known. The great project of the Renaissance, the surveying of the entire planet, would soon be accomplished.

Americans learn about this event as "the closing of the frontier" (famously described as such by Frederick Jackson Turner in an 1893 lecture to the American Historical Association). It was not just the American frontier that would soon be gone, however. Everywhere around the globe, lightly populated lands were being brought under the jurisdiction of markets, technological systems, and political oversight. This process involved more than a "scramble for Africa." It brought a speedy and final end to the independent societies of indigenous peoples all over the world, in a relentless project of genocide, dispossession, and domination based on an overwhelming superiority in firearms. The mapping of the planet coincided with a global invasion of Western civilization, so-called.

But if mapping and conquering the planet were regarded as a great triumph by many Westerners, it was also a source of anguish for them. Henceforth there would be precious little space available for adventure, discard, refuge, or experimentation. More and more would the planet be dominated by human desires, needs, and actions. The end of the age of exploration opened a new age of geography, when the world became both known and closed.[6]

This particular event of consciousness is the context—not a cause, but an important element of historical context—for the emergence of "design" as a key concept. Design is something humans have practiced since the Neolithic Era. (Indeed, the very concept of a "New Stone Age" is based on the design of stone tools.) The end of the world frontier, however, raised consciousness of the opportunities and necessities of better organizing the known world to fit human needs and desires. What is new in the late 19th century is the emergence of design as a conscious project, with articulated goals and methods, offering new ways to explore and take control in a new habitat.

As the age of exploration was drawing to a close, both the word and concept of *design* changed dramatically. Before then, according to the Oxford English Dictionary, neither as a verb nor as a noun did design have anything like its current meaning. From its first appearance as a verb around 1400, design meant appointing, nominating, or designating. Later, the verb could refer to planning or intending, but only in the human mind. Its usage as a noun appeared in the late 1500s, but again referred to a plan conceived in mind only. Associations of the word with a sketch, drawing, or physical plan did not emerge until the late 1600s. And only in the mid-to-late 19th century did design assume its current meaning, encompassing the arts of drawing or sketching, the processes of modeling and constructing, and otherwise planning features of an object in conformity with aesthetic or functional criteria.

At the same time new professions, arts, manufacturers, and markets emerged that proclaimed themselves organized around design. This sequence is especially notable in Britain, where there is a clear and striking series of episodes: the role of Owen Jones in the decoration of the Great Exhibition of 1851 in Hyde Park, London; the influence of John Ruskin; the work and ideas of the Pre-Raphaelites; the career of William Morris; and the founding of both the Century Guild of Artists and the Arts and Crafts Exhibition Society. Toward the end of the 19th century this stream of events broadened into the Arts and Crafts movement, and eventually Art Nouveau, as practices, ideals, and forms spread to the European continent, the Americas, and beyond.

This is not the place to follow this story into the 20th and 21st centuries. It is enough here to point out that the rise of design consciousness has continued to expand and intensify to this day. One clear indication of this development would surely be the founding of the Harvard Graduate School of Design under that name in 1936. Another might be the even more recent convergence of engineering and design education at the Massachusetts Institute of Technology, which in 2016 initiated a new Minor in Design, and which has, in the words of its current president, "taken steps to promote design thinking across the Institute through the lens of problem *setting*."[7]

What is consistent throughout this history is the tone of celebration for the brave new world of design. At a time when the closing of geographical frontiers aroused an uneasy mix of triumphalism and anxiety, design seemed to offer new kinds of frontiers, new openings for human discovery and mastery. This tone is even expressed in the statement of purpose for this journal, *New Geographies*, which begins:

> The journal … aims to examine the emergence of the geographic—a new but for the most part latent paradigm in design today—to articulate it and bring it to bear effectively on the agency of design.… It is time to consider the expanded agency of the designer.[8]

The editors of *New Geographies* want to investigate how "design practices can have a more active and transformative impact on the forces that shape contemporary urban realities."[9]

Thinking of design in such ambitious terms comes from the awareness that human domination of the planet requires and offers a much greater role for human design. What challenges, and what opportunities! But here too the sense of triumph is muted by a sense of anxiety, even anguish. As the scope of design expands and evolves, it inevitably reduces the availability of the un-designed world as a source of inspiration and knowledge. Inevitably, too, the expanded agency of the designer may prove unmatched to the power of historical forces such as those that "shape contemporary urban realities."

Let us look at some of the complexities of this new design consciousness by focusing on William Morris (1834–1896), arguably the individual most responsible for the rise

---

6  Rosalind Williams, *The Triumph of Human Empire* (Chicago and London: University of Chicago Press, 2013): ix–xi, 9–11. The title of the highly regarded work of Paul N. Edwards, *The Closed World* (Cambridge, MA: MIT Press, 1988) is echoed here, though as its subtitle (*Computers and the Politics of Discourse in Cold War America*) suggests, this book does not focus on the events I am describing here.

7  MIT President Rafael Reif, "Design at MIT: Inventing Excellent Answers," *Spectrum* (Winter 2017): 1. In January 2015 Hashim Sarkis, professor at the Harvard University Graduate School of Design, was chosen as dean of MIT's School of Architecture and Planning.

8  *New Geographies* Journal statement of purpose, initially written by the founders of the journal in 2009. www.new-geographies.squarespace.com/about.

9  Ibid.

of design in the 19th century, both in theory and in practice. No one presents a broader, deeper vision for the future of design—and no one better understands its limits.

\*\*\*

In 1883, when Morris filled out his membership card for the Democratic Federation (later the Socialist Democratic Federation), he identified himself in one word: "Designer."[10] Two decades earlier, upon graduating from Oxford, he had taken an apprenticeship in an architectural firm. It lasted only nine months, partly because Morris was temperamentally unsuited to the tedium of office work, and partly because he had greater ambitions. As a student Morris had discovered the writings of John Ruskin. Toward the end of a revelatory tour of Gothic cathedrals in northern France, on the docks of Le Havre, Morris and his friend and travel companion Edward Burne-Jones vowed to dedicate themselves to "a life of art."

For Burne-Jones, art meant painting. For Morris, it was not so clear what his art would be. Over time, under the influence of Ruskin, the Pre-Raphaelite brotherhood, and his own defiant spirit, Morris was increasingly drawn to the "lesser arts" of decoration, committing himself to narrowing the distinction between them and the fine arts. He would later write, in an essay titled "The Lesser Arts," that he wanted to "help bring forth decorative, noble, *popular* art." Such art would sanctify the human-built world:

> That art will make our streets as beautiful as the woods, as elevating as the mountain-sides: it will be a pleasure and a rest, and not a weight upon the spirits to come from the open country into a town…[11]

The artistic brotherhood slowly evolved into a creative collective. In 1861, along with six others from this group, Morris founded a decorative arts company that became known simply as The Firm. At the outset, The Firm primarily executed commissions for wall paintings and papers, stained glass, metalwork, jewelry, sculpture, embroidery, and furniture. The painters among the group were commissioned to produce drawings for stained-glass windows or tapestries. Morris's architect friend Philip Webb designed simple, sturdy furniture. The Firm used flexible production methods for a specialty market. Some items were made by hand; others were produced by machinery in The Firm's workshops or were outsourced to other manufacturers.

As business grew, The Firm moved to various locations in and around London. Eventually, in 1875, Morris took over sole management of what became Morris & Co. He was the epitome of the hands-on manager. In developing new lines of business, and also for his own pleasure, Morris took up one craft after another, teaching himself oil painting, calligraphy, embroidery, and illumination, among other arts. As a more or less conventional designer, his strongest talent lay in creating patterns for two-dimensional products like fabrics, carpets, and wallpapers.[12] He would hand-sketch the patterns, adding marginal directions for the colors of each detail, and identifying threads or pigments by number.

But for Morris, the work of design extended to all stages of the manufacturing process. Since most of the fabric and paper production was outsourced, he worked closely with manufacturers to ensure that dyes were true and correctly adjusted to the different materials being printed. He became interested in dyeing wools for tapestries and carpets. He researched the water flow around mill sites to be certain it was adequate for handling the dyeing vats as well as running the machinery. Morris became intrigued by traditional weaving techniques displaced by mass production methods. He eventually installed a loom in the bedroom of his London home, where he enjoyed the rhythmic labor of hand weaving.

Morris's deepest conviction was that the work of design includes the design of the labor process. When he came out publicly as a socialist in an 1883 lecture at Oxford, Morris stressed that "art is man's expression of his joy in labor." (Ruskin, who chaired the event, was unable to disguise his shock at hearing Morris declare himself a revolutionary.) When the lecture was printed as a pamphlet, Morris's assertion was set forth boldly, on a separate line, in capital letters.[13] The enemy of joy in labor, Morris declared, was not machinery but class division: not the "tangible steel and brass machine … but the great intangible machine of commercial tyranny, which oppresses the lives of all of us."[14] The Firm represents his design for a business that will bring some pleasure to worker and consumer alike within the context of the greater machine of capitalism.

For Morris there is no contradiction between being an owner-manufacturer and being a socialist. But there is a contradiction between the "desire to produce beautiful things," which engaged his work as a designer-manufacturer, and his two other strongest desires: to be closely connected with nonhuman nature and with human history. This tension is expressed most clearly in one of his best-known essays, "How I Became a Socialist," written in 1894 at the request of the editor of the periodical *Justice*. In it Morris declares, "Apart from the desire to produce beautiful things, the leading passion of my life has been and is hatred of modern civilization." He then describes how downcast he felt before his "conversion" to socialism because modern civilization seems to destroy the things he cares about most: "a deep love of the earth and the life on it, and a passion for the history of the past of mankind."[15]

The "beautiful things" that Morris designed were intended to provide connection, for him and for others, to the life of the earth and human history. But no matter how much he resists the oncoming tide of "modern civilization," the tide keeps coming: it is stronger than the designs are. Over and

---

10   Graeme Sutherland, "William Morris, Designer," insert in *William Morris: News from Nowhere and Selected Writings and Designs*, ed. Asa Briggs (1984; repr., London: Penguin, 1962), n.p.

11   William Morris, *The Collected Work of William Morris*, ed. May Morris, vol. 22 (London: Longman, Green, 1910–15): 26–27.

12   Williams, *Retooling*, 154–57.

13   Williams, *Retooling*, 190–91. Examples of these drawings can be found in Sutherland, "William Morris, Designer."

14   Sutherland, "William Morris, Designer," n.p.

15   Briggs, ed., *William Morris: News from Nowhere*, 33; for the passage containing the word "conversion": 35–36.

over again, in letters and in talks, Morris expresses his dismay at the junky buildings encroaching everywhere; cathedrals "restored" in false and ugly ways; forests and rivers trimmed and channeled into well-manicured parks; wonderful stories of human heroism and devotion lost or forgotten; lovely cities like Oxford invaded by commercial development. While Morris was not conventionally religious, he saw the world and the past as sacred precincts constantly being profaned.[16]

Worse yet, to some extent the designer is complicit in this process. His or her designs may be beautiful, but they still play a role in the constant extension of human presence, in the constant acceleration of human conquest of the planet. They are inherently, unavoidably, associated with humankind and the present, rather than nonhuman nature and the past. Just as mapping the world devours the unknown, so does the activity of design devour the un-designed, which Morris continued to value as a resource for art, and as a source of consolation and perspective.

Biographer E. P. Thompson analyzes Morris's turn to socialism as a means to reduce his own anxiety, depression, and even despair that had accrued over the years. Upon joining the Democratic Federation in 1883, Morris wrote to a fellow Socialist:

> I have not failed to be conscious that the art I have been helping to produce would fall with the death of a few of us who really care about it, that a reform in art which is founded on individualism must perish with the individuals who have set it going.[17]

Morris spent most of his time speaking, writing, and organizing, with the firm conviction that socialism would shape the forces of history in ways that design alone was incapable of doing. By the later 1880s, however, his energies were increasingly consumed by internecine quarrels on the British Left. Morris always remained a socialist, but came to realize that the revolution would not come within his lifetime, and maybe not for a long time afterwards.

Thompson concludes that in Morris, "we may see . . . not a late Victorian, nor even a 'contemporary,' but a new kind of sensibility."[18] It is a sensibility shaped by the triumph of human empire, where delight in new mastery over the planet is attended by endless experiences of change and loss. Human connections with nature and history are always fragile. The expectation that the world provide a stable, predictable platform for design cannot be taken for granted. Morris was pugnacious and energetic, but he lived with a constant sense of loss and foreboding. And he was bravely realistic in accepting the fact, as he saw it, that he was fighting a losing battle.

The example of Morris suggests that design is not just a new profession, a new way of transforming the world through art, but also a new consciousness: one that revels in the broad new possibilities for human powers, but that is also constantly sensible of the prospects for loss and of the limits to design.

To have the whole planet, and maybe even some of the cosmos beyond it, available to design can seem wonderful or frightening or both. Is the fully known world necessarily a welcoming one for human life? Is the fully designed world one that is truly comfortable for us? Can design be redesigned to address loss and instability, or does it inevitably—if unintentionally—contribute to a sense of entrapment in "modern civilization"?

Morris himself responded to such questions by extending his design activities in new directions. In the spring of 1890, when anarchists and other factions drove his group out of the Socialist League, Morris responded by finding two new outlets for his "life of art." He shifted the focus of his design and manufacturing activities to bookmaking, which eventually became a new business venture, the Kelmscott Press. He also started writing stories about the deep historical origins of just, equitable societies. The two projects merged when in 1891 he used the press to publish one of those tales, *The Story of the Glittering Plain*, which had already appeared in a magazine, making it into a beautiful book.

This turn to storytelling was by no means sudden or surprising. From the time he entered university, Morris was what we would now call a folklorist. He retrieved, studied, translated, and disseminated romances, epics, and sagas from European and non-European sources. Eventually some of these tales formed the basis of *The Earthly Paradise*, a long poem published in four parts between 1868 and 1870. It was a bestseller and over time became staple reading material of the English Victorian bourgeoisie.

Morris himself moved on to favor folktales that he considered more closely aligned with the perspective of ordinary people . He started reading Nordic sagas, first in translation and later in the original language in collaboration with an Icelandic scholar. He immediately took to their "delightful freshness and independence of thought . . . the air of freedom . . . their worship of courage (the great virtue of the human race), their utter unconventionality . . ."[19] Inspired by these sagas, Morris began to write stories about early Europe, experimenting with various mixtures of prose and poetry. He wanted to go back in time to Europe as it was before the rise of the Roman Empire, to retrieve the sense of freedom he imagined that the European tribes enjoyed when they still governed themselves. Morris also went into the future in his imagination, writing a utopian prose romance, *News from Nowhere* (1890), about a society that has rediscovered and affirmed its connections to the life of the earth and to the human past.

While writing *News from Nowhere*, Morris was also working on another project, a sort of fairy tale. He wrote to his wife, "I have begun another story but do not intend to hurry it—I must have a story to write now as long as I live."[20] This project became *The Story of the Glittering Plain*, which one scholar has described as "simply unlike anything else in English literature before it."[21] In this work Morris designs a coherent, independent habitat that is earthlike but not identifiable as any place on earth, which presents an engaging field for action,

---

16   Williams, *Retooling*, 228–29.

17   Briggs, ed., *William Morris: News from Nowhere*, 32; from a letter to Andreas Scheu.

18   Williams, *Retooling*, 227.

19   Briggs, ed., *William Morris: News from Nowhere*, 31; from a letter to Andreas Scheu.

20   Williams, *Retooling*, 215.

21   Richard Mathews, *Worlds Beyond the World: The Fantastic Vision of William Morris* (San Bernardino, CA: Borgo Press, 1978): 34–35.

learning, and heroism. In subsequent stories, Morris invented a whole series of new geographies. The importance of the imagined boundaries between land and water are evident in their titles: *The Well at World's End* (1896), *The Water of the Wondrous Isles* (1897), *The Wood beyond the World* (1894), and *The Sundering Flood* (1897).

These stories introduce alternative worlds that can be inhabited by a reader, viewer, or player. In the next generation, J. R. R. Tolkien, deeply influenced by Morris as linguist, folklorist, and writer, would name these inventions Secondary Worlds. Today they are called fantasies, and Morris is respected as a pioneer fantasist. Yet for Morris, these works present realistic visions of what human life could and should be like. They provide a setting for how humans should live, when the actually existing world of "modern civilization" does not.[22]

This synthesis of two kinds of design—of Secondary Worlds, and of splendidly crafted books—occupied Morris from 1890 until his death in 1896. Although he did not see the socialist revolution on the near horizon, neither was he trapped by history. The worlds and stories he invented offered a sense of human life as it should be, of experience liberated, of reprieve from the desecrations that he found so oppressive in "modern civilization." "William Morris: Designer" redesigned design to include new geographies.

\*\*\*

Design, as we know it today, is characterized by beliefs in the agency of the designer and in the potential of design to transform the human-built world for the better.

William Morris shared these convictions, but he also had strong doubts about them. He expanded the concept of design to include the labor process as part of the design process. Later he became a revolutionary Socialist because he believed design practice, no matter how broadly defined, could not bring about the collective changes that human life deserved. In the last years of his life he turned to designing imaginary worlds as settings for adventurous stories. He did this to nourish himself and others with comfort, inspiration, and pleasure. He was no longer trying to change the world, but to enjoy and accept the world as it exists—a world, as he explained in a letter to a dear friend, that "goes on, beautiful and strange and dreadful and worshipful."[23]

Morris's brave realism—his coming to terms with the forces of modern civilization without illusion or evasion—provides a model for designers today who face political and environmental problems that seem and perhaps are insoluble. One response to these problems is to keep expanding the scope of design to new things and systems: more of the same, but with grander scale and scope. Morris's example suggests another response. Giving design the burden of changing history may be too heavy a load for it to bear. There may be other ways that design can make the world a better place, through rethinking what "design" is and does.

---

22  Williams, *The Triumph of Human Empire*, 223–27.

23  Quoted in Williams, *Retooling*, 195. From Fiona MacCarthy, *William Morris: A Life for Our Time* (New York: Alfred A. Knopf, 1995): 363–64. The letter was written to a beloved friend in 1876, but we do not know which one; it was never sent.

Astronaut Bruce McCandless floating free in space, 1984.

The wholeness of the image is familiar to us, but the particular quality of fragility it reveals is harder to see. As massive as our planet is (some 6 sextillion tons), the habitable environment of the globe, as revealed by our own fragility, is but a tissue-thin atmosphere. The inextricably interrelated quality of this atmospheric space, as well as the increasing urbanization that shapes it, are the inescapable issues of our own age. — Nicholas de Monchaux, 2011

# More-than-Human Constellations as Immuno-Biopolitical Fantasy in the Urbicene

# Erik Swyngedouw

# The Urbicene

This essay mobilizes "the Anthropocene"—or rather, the Urbicene—the now popular term denoting a new geological era in which humans have arguably acquired planetary geophysical agency. Earth scientists, who coined the term, now overwhelmingly understand the earth as a complex, nonlinear, and indeterminate system with multiple feedback loops and heterogeneous dynamics in which (some) human activities are an integral part of terraforming processes. Planetary urbanization is of course the geographical expression of this anthropocenic process. Therefore, Urbicene might be a more appropriate term to capture the sociomaterial form that the Anthropocene takes. Planetary urbanization is predicated on intensifying proliferations of metabolic vehicles, in the form of techno-natural intermediaries that etch the transformation of nonhuman "stuff" in myriad socio-ecological, metabolic, and cyborgian relations that reorder human/nonhuman assemblages in radically uneven ways, with profound implications. Consider, for example, how the purportedly dematerialized affective economies that animate much of contemporary urban social and cultural life (IT networks, social media, smart networks, eco-architecture, big data informatics, and the like) are dependent upon mobilizing a range of minerals like coltan (columbite–tantalite); upon feverish resource grabbing, often through tactics of dispossession in socio-ecologically vulnerable places; upon production chains that are shaped by uneven and often dehumanizing socio-ecological processes; and upon "recycling" programs that return much of the e-waste to the dystopian geographies of Mumbai's or Dhaka's informal suburban wastelands.

The capitalist form of planetary urbanization and the processes that animate its combined and uneven socio-ecological catastrophe on a world scale are now generally recognized as key drivers of anthropogenic climate change and other socio-environmental transformations such as biodiversity loss, soil erosion, the construction of large-scale infrastructures like dams and energy plants, deforestation, ocean acidification, mining and resource extraction, pollution, and the increasing commodification of all manner of natures. Our urban fate and nature's transformations are irrevocably bound up in an intimate, intensifyingly metabolic—and highly contentious—symbiosis. The configuration of this urban metabolic relationship has now been elevated to the dignity of global public concern as it has become clear that deteriorating socio-ecological conditions may jeopardize the continuation of civilization as we know it.

Indeed, a global urban intellectual and professional technocracy has spurred a frantic search for a "smart" socio-ecological urbanity; prioritizing such qualities as eco-development, retrofitting, sustainable architecture, adaptive and resilient urban governance, the making of new interspecies ecotopes, the commodification of environmental "services," and innovative (but fundamentally market-conforming) eco-design. These techno-managerial dispositifs that suggest eco-prophylactic remedies for our predicament have entered the standard vocabulary of both government and private actors, and have now permeated the frontier of architectural, planning, and urban design theory and practice—fields presumably capable of saving both city and planet, while assuring that life on earth endures a little longer. Under the banner of radical techno-managerial restructuring, the focus is squarely on how to sustain capitalist urbanity so that nothing really has to change!

Given the desperate condition of the earth and the pivotal role of the urbanization process in this socio-ecological transformation, the city today is staged as experimental terrain for the implementation of new socio-technical arrangements that render it not only smart, but also "resilient" in the face of uncertainty and "adaptive" to potentially disruptive processes of rapid socio-ecological change. The proliferation of prophylactic socio-technical assemblages for urban resilience and sustainability coincided with the emergence of a radical ontological shift articulated around nonlinearity, complexity, contingency, risk, and uncertainty. Alongside, theorists from both the social sciences and the humanities proposed new earthly cosmologies, combining new materialist perspectives with more-than-human ontologies, that seek to grasp worldly matters in a more symmetrical human/nonhuman, if not posthuman, constellation. This contribution argues that the new materialist or posthuman ontologies—coincident with the new science of radical uncertainty, nonlinear dynamics, and complex, non-deterministic feedback loops—have become an integral part of the proliferating techno-managerial efforts to sustain and support the rapid pace of capitalist planetary urbanization. And further, they deepen the immuno-biopolitical governance arrangement that marks the present conjuncture. In this process, the matter of politics is fundamentally depoliticized.

## A More-Than-Human Ontology?

The Anthropocene argument not only recognizes the active role of humans in co-constructing the earth's geohistorical time, but also introduces potentially a new ontological frame of relational symmetry between humans and nonhumans. This horizontal relational ontology, variously referred to as more-than-human, posthuman, or object-oriented ontology, fuels the possibility of a new cosmology; that is, a new ordering of socio-natural relations.[1] Nonetheless, this new cosmology also has the potential to deepen particular capitalist forms of human-nonhuman entanglements, and can be corralled to support hyper-accelerationist, ecomodernist practices in which science, geoengineering, terraforming technologies, and big capital join to save both earth and earthlings.[2]

In the advent of the Anthropocene, geoscientists and earth systems experts discern the possibility, if not

---

1 See Diana H. Coole and Samantha Frost, eds., *New Materialisms: Ontology, Agency, and Politics* (Durham, NC: Duke University Press, 2010); Timothy Morton, *Hyperobjects: Philosophy and Ecology after the End of the World* (Minneapolis: University of Minnesota Press, 2013); Isabelle Stengers, *Cosmopolitiques* (Paris: La Découverte, 2003); Graham Harman, *Immaterialism: Objects and Social Theory* (Cambridge: Polity Press, 2016).

2 Here drawing on the work of Frédéric Neyrat, *La Part Inconstructible de la Terre* (Paris: Editions du Seuil, 2016).

necessity, for the management and careful "adaptive" and "resilient" massaging of the total earth system. The recognition of the earth as an intricately intertwined socio-natural constellation indeed opens up the possibility that—with loving supervision,
intelligent crafting, adaptive techno-natural machinery, and subtle manicuring—it can be terraformed in particular and resilient ways. As Bruce Braun has argued, in his careful dissection of the new materialisms, the parallels between non-deterministic geoscience, resilience studies, and the varieties of new materialisms associated with a more-than-human ontology are not difficult to discern.[3]

While previously the nonhuman was considered recalcitrant, uncooperative, and even prone to revengeful action when subordinated and marshaled for capital's use, a symmetrical ontology permits, at least in discourse and in the imaginary, a potentially more benign and modest, mutually supporting, sustainable, adaptive, and resilient assembly of human-nonhuman relations—conditions which, with some manipulation, in fact encourage capitalism to deepen further socio-ecological imbrication, while recognizing both the acting and incalculability of the nonhuman. The desire to produce a new terrestrial configuration unfolds through signifiers like "earth system governance" and "planetary stewardship, and is translated on to the urban terrain as "smart" urbanism, and "resilient" city planning.[4]

In this staging of the "good" Anthropocene, the new symmetrical relational ontology functions as a philosophical quilt that sustains and advocates accelerationist manifestos for hyper-modernization.[5] To save the world and ourselves, it seems, we do not need less capitalism, but rather a deeper, more intense, and radically reflexive form that revolves around reconstructing DNA and genetic material, harnessing nuclear power to drive the economy, forcing gas out of shale formations so that it can be "carbon-stored" elsewhere, and otherwise terraforming the planet such that capitalism can survive within a transformed earth. All this is supported by the monitoring and analysis of big data in real time. By covering up the contradictions of capitalist ecomodernization, the apparently revolutionary new material ontologies offer new storylines, and new symbolizations of the earth's past and future that can perform the necessary ideological groundwork. When this perspective enters the field of politics—the governing of things and people in common—it does so in a troublingly depoliticized manner.

# The Depoliticized Politics of the Anthropocene as Immuno-Biopolitical Fantasy

As suggested above, some anthropocenic narratives provide for an apparently immunological prophylactic against the threat of an irredeemably external and revengeful nature. In what other ways can the mainstreaming of critical and supposedly radical new ontologies—whose explicit objective is the unsettling of modernist cosmologies—be understood? And how can this Anthropocene's god trick be accounted for? Roberto Esposito's analysis of biopolitical governmentality, enhanced by Frédéric Neyrat's psychoanalytical interpretation, may begin to shed some light on this development.[6] Esposito's main claim expands on Michel Foucault's notion of biopolitical governmentality as the quintessential form of modern liberal state governance by demonstrating how it is increasingly supplemented by an immunological drive—with the mission to seal off objects of government (the population) from possibly harmful intruders and destabilizing outsiders that threaten the bio-happiness, if not sheer survival, of the population; thus guaranteeing that life can continue to be lived. This immunological drive is predicated on the suspension of communal gift giving, a form of asylum that suspends the call to participate in the rights and obligations of the commons: the community. The neoliberal injunction to enjoy individual freedom and choice is precisely the foundation of such an immuno-biopolitics; that is, the ring-fencing of fragmented bodies from interaction with the rights, obligations, and violence bound up in shared community life.[7]

Immuno-biopolitics are clearly at work, for example, in hegemonic Western institutional practices governing immigration, health, and international terrorism. A rapidly expanding arsenal of soft and hard technologies have been put in place in an ever more dense layering of technical, infrastructural, digital, and institutional-legal dispositifs—from tighter immigration laws, big data profiling systems, and continuous surveillance to the actual construction of steel and concrete walls and barriers, and the proliferation of all sorts of detainment camps and other militarized enclosures. Similar examples can be identified in the strict quarantines that arise when infectious diseases threaten to spatialize in ways that could penetrate the immuno-engineered ecotopian bubbles of the elite. Is it not the case that much of the urban sustainability, resilience, and techno-managerial practices that pervade ecological interventions, smart technologies, and governance

---

[3] Bruce Braun, "New Materialisms and Neoliberal Natures," *Antipode* 47:1 (2015): 1–14. See also John Protevi, *Life, War, Earth: Deleuze and the Sciences* (Minneapolis: University of Minnesota Press, 2013).

[4] Frank Biermann, "'Earth System Governance' as a Crosscutting Theme of Global Change Research," *Global Environmental Change* 17, no. 1 (February 2007): 326–37; Will Steffen, Asa Persson, Lisa Deutsch, Jan Zalasiewicz, Mark Williams, Katherine Richardson, Carole Crumley, Paul Crutzen, Carl Folke, Line Gordon, Mario Molina, Veerabhadran Ramanathan, Johan Rockstrom, Marten Scheffer, Hans Joachim Schellnhuber, and Uno Svedin, "The Anthropocene: From Global Change to Planetary Stewardship," *AMBIO: A Journal of the Human Environment* 40, no. 7 (2011): 739–61.

[5] See Clive Hamilton, Christophe Bonneuil, and Francois Gemenne, eds. *The Anthropocene and the Global Environmental Crisis: Rethinking Modernity in a New Epoch* (New York: Routledge, 2015); Erle Ellis, "Planet of No Return: Human Resilience on an Artificial Earth," *The Breakthrough Journal* 2 (Fall 2011): 37–44; Frédéric Neyrat, "Critique du Géo-constructivisme Anthropocène & Géo-ingénierie," *Multitudes*, 56, http://www.multitudes.net/critique-du-geo-constructivisme-anthropocene-geo-ingenierie/.

[6] Roberto Esposito, *Bios: Biopolitics and Philosophy* (Minneapolis: University of Minnesota Press, 2008) and *Immunitas* (Cambridge: Polity Press, 2011); Frédéric Neyrat, "The Birth of Immunopolitics," *Parrhesia* 10 (2010): 31–38.

[7] Alain Brossat, *La Démocratie Immunitaire* (Paris: La Dispute, 2003).

are precisely aimed at reinforcing the immune system of the body politic against threatening outsiders (like $CO_2$, waste, bacteria, refugees, viruses, ozone, financial crises) so that life as we know it can go on? Immuno-biopolitics deepens biopolitical governance in an era of uncertainty and perpetual risk.[8] As Pierre-Olivier Garcia writes:

> An immunitary power takes control of the risks, dangers, and fragilities of individuals, to make them live in a peaceful manner while obscuring any form of dissensus.[9]

Roberto Esposito and Alain Brossat call this *immunitary democracy*. This social configuration operates as a system that guarantees untouchability; the sense of being immunized. It is a fantasy of total protection and the complete securitization of life, without exposure to risk. For Brossat, this is a dangerous fantasy, as the immunitary logic entails nothing other than the destruction of community, of being-in-common. By necessity, immunitary logic requires the continuous production of the exposed and the exiled (i.e., the non-immunized) alongside the immunized, and leads to de-politicization. The immunized become mere spectators of the suffering of others from the cocoon of their sanctuary eco-topian spaces, and from which only the affective registers of hatred or compassion remain as flipsides of the same depoliticizing process.[10] Of course, as Neyrat insists, the immunitary dispositif does not actually function in this way, because exposure to risk affects all—though not to the same extent.

In relation to refugees, (bio-)security, and the economy, the immuno-biopolitical gesture occasionally succeeds in trans-locating risk and the fear of acquiescence (while nurturing them all the same) onto the terrain of a crisis to be governed and managed, or a situation to adapt to. But this immuno-biopolitical dispositif of crisis management is rapidly disintegrating in the face of uneven, actually existing socio-ecological catastrophe. Indeed, with respect to our socio-ecological condition, the standard technologies of neoliberal governance—which nurture the immuno-biopolitical desire that is paradoxically also, Esposito points out, the primary logic of neoliberal governmentality—become increasingly ineffective. Few believe, for example, that the 1.5 Celsius limit on global temperature rise, set by the "international community," will be achieved. Is it not the case that the immuno-biopolitical managerial tactics of earth system governance, geoengineering, urban adaptation technologies, and other eco-governance arrangements leave an uncanny remainder? Although our ecological condition enjoys the dignity of global public concern, are we not left with the gnawing feeling that the socio-ecological parameters keep eroding? Although a combination of market-led adaptation and mitigation strategies are implemented as a firewall against dangerous climate change and other eco-calamities, the hard reality of ecological disintegration continues apace. While other risks (economic or geopolitical/security) are mollified by immuno-biopolitical gestures that promise life unencumbered—for the protected, thereby (re-)producing and expanding the exposed—in the face of potentially lethal threats by deepening immunological management, screening, and techno-shielding, the environmental biopolitical masquerade secures at best temporary relief. The ecological and climate conundrum obscures consequences from us even as it emanates real danger, both for the immunized and the exposed, no matter the range of palliatives at our disposal. The woefully inadequate and failed attempts to immunize life from the threat of ecological collapse cannot any longer be ignored.

From the Lacanian psychoanalytic perspective, the Real of socio-ecological destruction insistently intrudes on this immunological fantasy script, exposing its unstable core, uncovering the gap between the Symbolic and the Real, undermining its supporting imaginary matrix and thereby threatening the coherence of the prevalent order. This incessant return of the Real may fatally subvert our drive's (in a psychoanalytic sense) primordial energy as we are increasingly caught up in the horrifying vortex of radical and irreversible socio-ecological disintegration. The human fantasy of eternal life meets the certainty of its inevitable and always premature end. A radical reimagination of the earth system, barring the untimely death that is now firmly on the horizon, is urgently necessary. But instead, the uncanny feeling that all is not as it should be is sublimated and objectified in *objet a*—the horrifying "thing" around which both fear and desire circulate. It is a fear vested in an ambiguous "outsider" that threatens the coherence and unity of our life-world.

As Esposito argues, the immuno-biopolitical dispositif becomes displaced in a thanatopolitics of who should live or die. Eventually, the excessive immunological drive turns against that which it should protect and becomes self-destructive in a process of autoimmunization. For example, the construction of urban eco-bubbles for the privileged simultaneously produces unprotected exiles and deepens ecological destruction elsewhere. This is eco-gentrification at its height. The very mechanisms that permitted biopolitical governance in the 20th century—the thermocene of unbridled carbon metabolization and energy production—soon developed an autodestructive process of isolating pathological syndromes as externalized bad agents in need of sequestration.[11] In other words, the mechanisms allowing the production of secure life end up threatening its very continuation. This infernal dialectic, Neyrat contends, is predicated on redoubling the fantasy of absolute immunization: despite the fact that we know very well we will die, we act and organize life as if it will go on forever.[12] When the excessive action of an external threat—such as $CO_2$—can no longer be ignored, the immuno-biopolitical drive broadens and intensifies. It proceeds through an ontological reversal: internalizing the pathological outsider so as to

---

8   Frédéric Neyrat, *Biopolitique des Catastrophes* (Paris: Les Prairies Ordinaires, 2008).

9   Pierre-Olivier Garcia, "Sous l'Adaptation, l'Immunité: Etude sur le Discours de l'Adaptation au Changement Climatique" (PhD diss., Université Grenoble Alpes, 2015), 321 (my translation).

10   Maria Kaika, "Between Compassion and Racism: Europe's New Janus-Faced Citizen, Or: How Fighting for the Commons Can Avoid an Anthropological Catastrophe" (paper presented at the annual meeting of the Association of American Geographers, Chicago, Ill, April 21–25, 2015).

11   Garcia, "Sous l'Adaptation, l'Immunité," 253–255.

12   Frédéric Neyrat, interview by Elizabeth Johnson and David Johnson, "The Political Unconscious of the Anthropocene," *Society and Space* (March 20, 2014), http://societyandspace.org/2014/03/20/on-8/.

render it "governable" while redoubling the phantasmagoria of absolute immunization, to the point of thanato-political self-destruction. Sustained by human exceptionalism (as the sole species capable of preventing its own death), both the modest and more radically accelerationist geo-imaginaries of the Urbicene find, in this fantasy space, ultimate ground.[13]

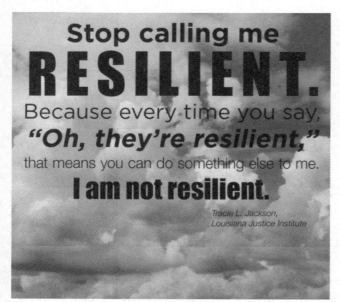

"Stop calling me resilient,"
Mural in the Lower Shankill, Ireland, 2015.

The symmetrical human-nonhuman foundation (as compared to the nature-culture ontological split of yesteryear) on which the Anthropocene rests promises to resolve the unbearable deadlock between immuno-biopolitics and thanato-politics without really having to alter the trajectory of socio-ecological change. In fact, the deadlock intensifies. In psychoanalytical terms, the immuno-biopolitical prophylactic proposed by the "good" Anthropocene circulates around the death drive—the obsessive pursuit of desire that shrouds the inevitability of "death." The immuno-biopolitical fantasy assures us that we can go on living without staring the reality of eventual (ex-)termination in the eye. While the pursuit of happiness lies in avoiding pain, the death drive—sustained by desire—propels us forward as if we will live forever, irrespective of the threats we encounter on our journey to the end. The energy of the drive is fueled by the disavowal of a certain death. It is the hysterical position that guarantees that death remains obscure and distant, an obscene impossibility. Becoming "resilient" is indeed the ultimate hysterical act, to which the only sane response is: "Stop calling me resilient!"[14]

The anthropocenic promise of a geoengineered earth or a more modest, caring society-nature relationship—the fulfillment of a constructivist materialist ontology— brings the incoherent whole of the earth within an immuno-biopolitical framework that promises, if not guarantees, the sustainability of civilization. And it is the urban where this techno-machinery is experimentally introduced and implemented. The outsider that threatens the integrity of our socio-spatial matrix (i.e., nature as we knew it) is captured by an ontology that allows for total incorporation. Such phantasmagoric staging of the Anthropocene depoliticizes the matter of nature. We can survive and do so without needing to face political action or confronting radically different political choices. A shift in the techno-managerial apparatus will suffice.

The immuno-biopolitical gesture confidently projects our survival without considering the need for a transformation of socio-natural relations themselves; it invites techno-managerial adaptations to assure the "sustainability" of the existing. In its ecomodernizing straitjacket, the Anthropocene (or Urbicene) offers a radical reinterpretation so that nothing really has to change. It promises to craft a manageable earth and guarantees our survival, freed from both uncertainty (despite endorsing reflexive consideration of the associated risks) and the destructive action of an external nature that is radically other than the human.

## Recentering the Political

While the controversies of the Anthropocene can be marshaled in many ways, suggesting indeed that the stuff of "things" is politicized, the political cannot and should not be grounded on the eventual truth of the Anthropocene. There is no code, injunction, or ontology that can (or should) found a new political ecology. The ultimate depoliticizing gesture resides precisely in letting the naming of a geosocial epoch decide our politics, thereby disavowing that the "our" or "the human" does not exist. The Anthropocene's cosmology is yet another failed attempt to establish a new politics on a contested truth of nature. Surprisingly, post-foundational political thought is rarely articulated with more-than-human ontologies. Although the post-foundational intellectual landscape brought complexity theory and the new materialisms into conversation, claims by the latter to offer radical new possibilities in fact fail to engage with concomitant post-foundational political thought. Jacques Rancière, for example, understands the political as the interruptive staging of equality by the "part of no-part."[15] For Rancière, the political appears when those who are not counted within the count of the situation (the excluded and exiled) make themselves heard and seen—that is, perceptible and countable—by staging equality. In post-foundational thought, the political emerges symptomatically as an immanent practice of appearance. For Badiou, it takes place as an *event* that interrupts a given relational configuration.[16] This performative perspective of politics needs no grounding in any current or historical order or logic, based on, say, nature, race, class, ability,

---

13  Neyrat, "The Political Unconscious."

14  Maria Kaika, "'Don't call me resilient again!': The New Urban Agenda as Immunology, Or: What Happens When Communities Refuse to Be Vaccinated with 'Smart Cities' and Indicators," *Environment and Urbanization* 29, no. 1 (April 2017): 89–102.

15  Jacques Rancière, *Disagreement: Politics and Philosophy* (Minneapolis: University of Minnesota Press, 1998).

16  Alain Badiou, *Being and Event*, trans. Oliver Feltham (London: Continuum, 2007).

or gender; the political is the ability and performative act to disrupt, disturb, and reconfigure what is perceptible, sensible, and countable.[17] This notion of the political coincides with the polis as a political community; the theater through which political space is carved out and where new forms of living together are experimentally enacted. A political-ecological community, therefore, insists on both community's *necessity and impossibility* as a fully closed, harmonious, internally coherent, and cohesive configuration.[18] Such a performative perspective focuses on excessive actions that trespass, undermine, and exceed existing situations and relational configurations.

Indeed, a wide range of political theorists, despite often radically opposing views, share the aim to renew political thought in a post-foundational landscape; an environment in which the ontological—in thought, in social processes, and in the world's more-than-human material acting—is this vanishing horizon of inconsistency, radical heterogeneity, and incalculable immanence. Yet their views are far from homogeneous.[19] Alain Badiou, for example, insists that recent attempts to reestablish the political philosophically, are in fact integral to what he considers pervasive processes of depoliticization. For Badiou, "ecology is the new opium for the masses."[20] A reemergence of the political, he insists, requires militant action but above all resides in the fidelity to egalitarian political events that may realize a truth procedure. The transition to political truth procedure requires subjects faithful to a catalytic egalitarian event, who continue to aspire to its generalization and coming into being through sustained action and militant organization. It is a fidelity to the practical possibility of the coming community, but without any ultimate ontological guarantee in history, theory, nature, the Party, or the State. Such fidelity slowly and relentlessly yields a new socio-physical reality, often in the face of the most formidable repression and violence. And it requires commitment, sustained militant action, and painstaking organization in the lengthy process of emancipation. Finally, it necessitates abandoning the fear of failure, as fail we shall. The unpredictable indifference of the more-than-human guarantees just that.

# Endgame?

The Urbicene and its symbolic framing is ultimately structured around ecologies of fear that nurture and cultivate a reactionary political stance and advocate techno-managerial forms of geo-social intervention. It is indeed the urban process and its modes of governing that undergird the dynamics of the Anthropocene. This techno-managerial bias is an expression of the current process of post-politicization worldwide and the deepening of our post-democratic condition, intensified by distinct immuno-biopolitical gestures that are also perversely self-destructive. The claim outlined above about the Urbicene and its performativity in no way suggests ignoring the diverse, multiple, whimsical, contingent, and often volatile socio-ecological relations of which we are part. But there is an urgent need to question the legitimacy of all manner of socio-environmental politics, policies, and interventions in the name of a thoroughly symbolized and humanized nature. The Anthropocene forecloses the political as its imaginaries are constituted and hegemonized, disavowing democratic deliberation by erasing the spaces of agonistic encounter.[21] Rather than rely on new terminology, we must accept the extraordinary variability of natures, make "a wager" on what kind of natures we would wish to inhabit, choose politically between this rather than that nature, and accept that not all there is can be known. To the extent that there is an earthly politics, it will have to attest to the heterogeneities implicit in the social that have the capacity to destabilize any community; interrupt the order of the sensible; subtract from the relational frame; and—in doing so—forge new egalitarian human-nonhuman entanglements.

---

17   Jacques Rancière, *The Politics of Aesthetics: The Distribution of the Sensible*, trans. Gabriel Rockhill (London: Continuum, 2004).

18   Jean-Luc Nancy, *The Inoperative Community* (Minneapolis: University of Minnesota Press, 1991).

19   For a review, see Japhy Wilson and Erik Swyngedouw, eds., *The Post-Political and Its Discontents: Spaces of Depoliticization, Specters of Radical Politics* (Edinburgh: Edinburgh University Press, 2014).

20   "Live Badiou: Interview with Alain Badiou, Paris 2007," in Oliver Feltham, *Alain Badiou: Live Theory* (London: Continuum, 2008), 130–39.

21   Chantal Mouffe, *On the Political* (London: Routledge, 2005). See also Wilson and Swyngedouw, *The Post-Political and Its Discontents*.

**Image Credits**
22: Courtesy of Extramural Activity

Sikai road, Songjiang District in Shanghai, 2013.

I want to start with a simple fact, which astonishes me. Between 1900 and 1999, the United States consumed, according to a US Geological Survey, 4,500 million tons of cement. Between 2011 and 2013, China consumed nearly 45 percent more cement than the United States had consumed in the whole of the preceding century. That magnitude of spreading cement around is unprecedented. Those of us who live in the United States have seen plenty of cement used over our lifetimes. But what has happened to China is extraordinary. And you can just imagine what some of the environmental, political, and social consequences might be. So the question I want to ask is: Why did this happen? — David Harvey, 2016

# Geographies of Sensitive Matter: On Artificial Intelligence at Urban Scale

**Benjamin H. Bratton**

# I. Matter

When considering "geography" as a design project, we need materialisms that are closely tuned to the philosophical provocations of actual matter. We know, for example, more about matter's programmability than when many now-conventional materialisms were first conceived, and this has profound implications, not only for how we understand matter, but also for how we understand ourselves *as matter*.

Toward this end, a question worth asking and re-asking is precisely how matter sometimes assembles itself into certain forms that can exhibit "intelligence." Such assemblages include the frontal cortex of *Homo sapiens*, but also non-carbon formats of intelligence, such as contemporary Artificial Intelligence (AI). Considering its mineral materiality, this paper begins to delineate a *geology* of AI. It is the basis of what synthetic intelligence at an urban scale is, and will be.

Infused and interwoven with ubiquitous computational infrastructure, urban life coheres in the aggregation of encounters and relations between people, things, and material events, including and perhaps especially those in which people are not directly involved. But even while so much is seen and registered, not everything sees—or looks back at what it sees—in the same way.

A bioeconomics enters; principles for the conservation of information and/or energy come into play. From gold standards to carbon pricing to the energy bill for Ether settlements, we are already confronted with fundamental questions about how matter and energy may contain, cohere, represent, guarantee, or exchange value. With blockchains or their derivatives, we may induce vernacular currencies that refer explicitly to the energy expenditure necessary to calculate them into existence (unlike dollars and yuan). To the extent that they are also provisionally programmable, we may soon have virtual currencies that address their only-apparent lack of materiality by indexing the weight of underlying hardware in the existential overhead of each token and transaction.

Put differently, energy may not be programmable, but currencies that represent energy expenditures can be. For an eco-economics, perhaps the key is to make the carbon chain more ubiquitous, not less, by expanding the scope of the sovereign blockchain wallet to include composite users, especially major non-hominid carbon sinks. Their landscape-scale agency is primary—not peripheral. As I argued in *The Stack: On Software and Sovereignty* (2016),

> part of the design question then has to do with interpreting the status of images of the world that are created by the program (as well as its image of itself) and the ways it governs a planet by governing its model of that planet. That model is built on the interrelationship of nonhuman biologies and chemistries as well, and so the images that the designer can deduce or produce serve the representational agency of matter that otherwise would be invisible to itself beyond its most local chemical interactions[1]

Such programs run deep, as computation becomes a solvent for algorithmic reason, already diffuse in matter. While the design implications are both profound and uncertain we can still make some interim conclusions. Starting from the premise that computation was more discovered than invented, we take it that computation is one of the ways that matter, in whatever form, achieves intelligence via procedural abstraction. Among these, sapient brain tissue is matter that has achieved intelligence in ways that both clearly are and clearly are not "computational" in any normal sense. The dissolution of synthetic computation into urban surfaces and fabrics provides them with enhanced forms of distributed intelligence, in addition to those they already have, because it provides new capacities for abstraction. That is, as generic principles, intelligence can be modeled as a capacity for abstraction, and computation is a technique of intelligence, so computation can model both procedural and aesthetic abstraction.

Such a perspective should recognize that intelligence is distributed among multiple positions and forms of life, both similar and dissimilar to one another. No single neuro-anatomical disposition has a monopoly on how to think intelligently. What might qualify as general intelligence is not duty bound to species or phylum, but is simply marked by the capacity for abstraction. Cognition, therefore, is truly distributed. The ability of an organism, however primitive, to map its own surroundings—particularly in relation to the basic terms of friend, food, or foe—is a primordial feat of abstraction from which we do not graduate, so much as learn to develop into greater accomplishments like reason and its local human variations. Thus cartographic abstractions are not merely an early stage through which things pass on their way toward more complex forms of intelligence, but rather form a core principle of that very complexification. As I wrote in the essay "Outing A.I.":

> In time, should complex AI arrive, it will not be humanlike unless we insist that it pretend to be so, because, one assumes, the idea that intelligence could be both real and inhuman at the same time is morally and psychologically intolerable. Instead of nurturing this bigotry, we would do better to allow that in our universe, "thinking" is much more diverse, even alien, than our own particular case. The real philosophical lessons of AI will have less to do with humans teaching machines how to think than with machines teaching humans a fuller and truer range of what thinking can be.[2]

---

Passages from this text were developed for various invited presentations, including the 2015 Premsela Lecture at Het Nieuwe Instituut, Amsterdam, and conferences at the Harvard Graduate School of Design in 2016, and Yale School of Architecture in 2015.

1  Benjamin H. Bratton, *The Stack: On Software and Sovereignty* (Cambridge, MA: MIT Press, 2016), 301.

2  Benjamin H. Bratton, "Outing AI: Beyond the Turing Test," *New York Times*, February 23, 2015.

## II. City

Even though they are software-based, AI phenomena are physical. Like protozoa and their cilia, feeling about to discover what is out there, or like humans seeing, tasting, and imagining patterns, today AI is (often) augmented by various technologies of machine cognition, machine sensation, and machine vision. These sensory apparatuses allow AI to see and sense the world "out there," and to realize forms of mechanically embodied intelligence, both deliberately programmed and emerging unexpectedly.

For urbanism, we do not begin with AI in a petri dish, or with AI as a kind of synthetic philosopher, but rather as a landscape-scale enterprise for which information sensing and processing exist in complex economic niches. Especially because AI has evolved in the back-and-forth between philosophical experiments and real technologies, AI's epistemic and urban careers cannot be disentangled, except under laboratory conditions. If intelligence is one way that matter organizes itself into a durable complexity, then a special form of that complexity is the city—that vast, settled accumulation of material intelligence, both human and inhuman. As artificial intelligence becomes more sophisticated and nuanced, what will be its urban future?

To understand that, we should conceive of AI as something we see around us all the time (and which preceded us) and also as technologies of synthetic algorithmic reason that we now design and designate. Both notions will help formulate cities as they intersect one another. We should see AI not in terms of "how we think that we think"—as a sort of virtual or artificial human cognition—or as *serving* us, but as something(s) that embodies another position than we do along a shared continuum of material intelligence.

Both forms of AI are fundamentally defined by sensing and sensation. With urban-scale sensing network arrays and their various attendant AIs, we mark only fuzzy boundaries between sensing the world and processing information about what is sensed. This condition is not ephemeral; it is prosaic. To consider the range by which animal, vegetable, and mineral matter "senses" its world does not require the projections of panpsychism, only attention to chemistry.

In many cases, the pairing of synthetic sensing (vision, for example) with evolutionary robotics allows for simple artificial species to behave intelligently because they have what amounts to a functionally embodied perceptual model of their habitats. This located heuristic knowledge is essential to how an AI manipulates a situated problem space. They can think because they can sense. The city—any city— is home to millions of species, from microbes to insects, to vegetation, to sapient mammals. It is a living bacteriological and immunological tumult, a situated ecology of predation and symbiosis, competitive and sexual selections, across multiple scales—into which we have now inserted landscape-scale machine sensing and evolutionary robotics.

The AI city may embody itself, but not as humans do. A commingling of diverse sensors (tracking light, air, sound, chemistry, etc.) sketches a garden of sensing and thinking little species, partially embodied discretely, one with another, and partially co-embodied with one another as their information inputs are aggregated, modeled, and acted upon in various pluralities.

*Homo sapiens* comes equipped with an extraordinary array of sensory faculties, which may be augmented further by synthetic layers in various combinations, from the sensors/trackers in the phones that we carry about like mules, to more intimate media of artificial images, sounds, etc. Situated in this expanded field, we are both sensors and sensed. On the one hand, we are the primary sapient actors in this drama, supervising an orchestra of technologies, each individually capable of functional processing and aggregate emergent intelligence (as are neurons). On the other, we are not only the subject of this scenario, but also its subject matter. The wider urban landscape of natural and synthetic sensory systems is not only a platform through which we extend and extrapolate our capacities for abstraction—as noted, it is also capable of feats of abstraction on its own. And so it is not just AI in the city, but nested AI co-embodied *as* the city, a different organism signaling to niche populations, to genome dynamics, and to structures at work.

We should thus imagine AI urbanism in terms of Jakob von Uexküll's "stroll" into the field populated by intermingling but mutually oblivious little lifeworlds, variously flying, flowering, digging, etc. In Uexküll's parable, the tick lies in wait for some threshold event to come its way, at which point it triggers a programmed response at, and into, its own void.[3]

Many of our urban sensors and their limited AI forms work similarly, and with similar nobility.

More versatile synthetic intelligences occupy more complex *Umwelten*—some are predator and prey, some are in motion, some are blooming, and some pollinating. As we stroll among them, we may be registered by these intelligences or we may be ignored. We may be a primary cause of concern, or we may be only passing interference in an evolutionary dynamic in which we are neither protagonist nor target.

The cities we build are in this way not only the habitat about which AI learns embodied contextual knowledge, they are also the distributed sensory apparatus with which the AI embodies that context. The city layer of The Stack—as a platform and as a governing apparatus—is self-incorporating in this way (and in others besides).

## III. Decoding

But while AI urbanism's synthetic systems may activate existing material chemistries in new ways, this does not just happen autonomously. It is a function of programmed matter, and so variants of code and coding are both in play and at work.

Decoding (and implicitly, coding) may entail many different design projects and problems. Some we are already

---

[3] Jakob von Uexküll, *A Foray Into the Worlds of Animals and Humans: With a Theory of Meaning*, trans. Joseph D. O'Neil (Minneapolis: University of Minnesota Press, 2013).

familiar with, others less so. One problem has to do with how software trespasses conventional distinctions between language and technology. That is, writing code in a software language, according to its syntactical norms, is a kind of linguistic competency.[4] But unlike natural languages, this code is executable. If Java is put in a machine, that machine will do what the code tells it to. Other than human machines, the same cannot (as of this writing) be said for English or Chinese language. So code, in this sense, is a language that is also a technology. But unlike mechanical technologies, it does things according to alphanumerically specific commands, which are dependent on hardware media, but are only physical commands at very granular levels.

The convertibility between software's linguistic and technical guises inspires some to apply critical literary or textual analysis to code itself. Decades ago, to "read the city"— decoding urban semiotics, unpacking the city-as-text—may have been the assignment. For many reasons, there has been a turn away from language, toward materialist perspectives, but not because language is somehow itself immaterial. Today instead of interpreting cleanly printed pages, we must account for ambient data shedding, surveillance, and encrypted signals (sent and received), so in fact the politics of decoding is based on uneven revelations of what has and has not ever been encoded. This is, like all economies of camouflage, a slippery game of optics, position and counter-position.

However, I prefer to foreground other connotations of coding/decoding for urban geographic design. Despite appearances, these may be more closely related to the legal codes that subdivide cities into discrete zones to preemptively organize activities and functions. But now the sovereignty of such decisions may reside less in the reach of the law than in locally executable software, embedded firmware, and infrastructural cloud platforms. Not only do we now ask software to do things that we once asked of architecture (such as organize circulation flow, and crowds in place), but we also ask it to do things that the law once did, no matter how it was imposed. To code and decode may mean understanding the interplay of these processes, as well as untangling them on behalf of another city yet to come. For some that means resistance through self-encapsulated possessive individualism. My entry point, however, does not prioritize negation, but rather sees code as there to be rewritten and overwritten. Code evolves and does so in relation to its contexts and its users, which include us, but which are not reducible to us only. For design, geographic materialism is based in a projective (as in project, and projectile) connotation of coding.

Toward this end, we focus attention on the back-and-forth between ambient computation as a niche in which users are the active organisms, and when users serve as the niche that exerts selective pressures on the code. We might trace one philosophical project for which the co-evolution of living beings and technical beings tracks phylogenetic lineages of technical objects in associations of mutual assimilation.[5] Our own psychological and social individuation is predicated on inheritances drawn from apparatuses that various ancestorsthave coaxed into being, and which continue to evolve in ways not reducible to their interactions with humans. Our urban conditions depend on these genetic and epigenetic adaptations, but for design theory in the age of machine intelligence, an inverse association is equally at work. The evolution of technology relies on its own phylogenetic extensions, which are often are other technologies, though sometimes we are the extensions. By prosthesis and inverse prosthesis, tools are not only our means of leapfrogging genetic evolution, we are also, in turn, a technology that technologies use to do the same.

Code at the level of the city may situate this dynamic in several possible ways. In the composition of urban-scale complex adaptive systems, software can be niche, signal, environment, agent, and/or boundary, etc. Even the same bit of code may do one or more of these at the same time: it may force a genetic or phylogenetic shift, and it may be the beneficiary or the victim of that forcing. It entails and encodes not only form but also force. This is why software works so well as a repository of what an early modern era termed "governmentality," even if another geometry of political geography is now at hand.

## IV. Gastronomy

None of this ensures that the "politics" of urban society responds in concert with what matters. When confronted, directly and indirectly, with the unpromising implications of ecological indifference to cultural traditions, hominids vote for sovereigns who pledge to bind them together somehow, and who claim powers to make reality obey the tribal narrative. By rituals of public voice, these arbitrary yarns are conjured by a magical politics of representation and identification. What Carl Schmitt called "political theology" comes into its own when the function of the state is to provide what amounts to a religious plotline. As publics and polities are swayed by regional myths and hemispherical *Grossräume*, we literally vote for mind over matter. Matter, however, is unconvinced.

In geographic materialism, what places are left for desire? *Eros* and *Thanatos* are tricky and probably deceptive tropes. It is not a sad fate to be interwoven with alien matter, and other anatomic and economic configurations are still possible. The particular economics of our current landscape of matter is only a default setting; it is not destiny. Chemistry may ultimately drive the most radical forms of political and philosophical imagination, perhaps toward a "culinary materialism."[6] Still, some insist that cities are smart only when they are filled with new software, and others that cities are only truly alive when synthetic algorithmic sensing and processing is absent, or silenced, or subordinated. Both perspectives make it too difficult to see our whether-we-want-to-or-not geodesign project for what it really is, or should be: a *molecular gastronomy at lansdscape scale*, a restored and resorted ecology designed to taste itself in new forms of richly spiced and imaginatively sauced mutual ingestion.

---

[4] Benjamin H. Bratton, "Can the Bot Speak?: The Paranoid Voice in Conversational UI" in *Across & Beyond: A Transmediale Reader on Post-Digital Practices, Concepts, and Institutions*, eds. Ryan Bishop, Kristoffer Gansing, Jussi Parikka, Elvia Wilk (Berlin: Sternberg Press, 2017).

[5] See, for example, Gilbert Simondon, *On the Mode of Existence of Technical Objects* (Minneapolis: University of Minnesota Press, 2016).

[6] A similar argument is developed concise by Robin Mackay and Reza Negarastani in their editorial introduction to *Collapse VII: Culinary Materialism* (June 2011): 3–37. See https://www.urbanomic.com/wp-content/uploads/2015/02/Collapse_7_Introduction.pdf.

Crescent Dunes, Nevada, 2016.

[T]he prospect of humans working in partnership with intelligent machines is not so much a usurpation of human right and responsibility as it is a further development in the construction of distributed cognition environments, a construction that has been ongoing for thousands of years. Also changed in this perspective is the relation of human subjectivity to its environment. No longer will the human be seen as the source from which emanates the mastery necessary to dominate and control the environment. Rather, the distributed cognition of the emergent human subject correlates with—in Bateson's phrase, becomes a metaphor for—the distributed cognitive system as a whole, in which "thinking" is done by both human and nonhuman actors.
— N. Katherine Hayles, 1999

# Against Nature:
# The Technological Consciousness of Architectural Design

# Luciana Parisi

In the last 15 years, the *generic* function of computation—or the rule-based processing of data behavior—has come to dominate the future image of architectural design. Its aesthetic dimension speaks directly of the technicity of contemporary culture, now corresponding to the machine vision of a world made of data. An inexhaustible archive of newly discoverable information is ready to be rescripted by powerful search engines whose sophisticated algorithms feed off the vast amount of increasingly fragmented and varied data.

This image rests upon a synthetic modality of computational functions and aesthetics, of procedural operations and spatiotemporal reconfigurations of perception and cognition. Far from merely an abstract application of the linear logic of algorithmic instructions to any set of data, this synthesis rather reveals the automated capacity to generate (directly design and print) increasingly particular levels of complex geometry. From data analytics to patterns of recognition, from generative to evolutionary algorithms, big data and machine learning, architectural design has already embraced this new conception of technicity—shifting between the parametric formalism of multi-agent behavior and the open synthesis of data's granular dimensions.

If, ever since the 1990s, architectural design has used computation to capture the physical movement of matter through generative algorithms that could render the point at which a line bends into a curve, then with the emergence of parametricism within architectural design in the 2000s, contingent values (e.g., atmospheric, geological, biological, physical variables) have been formally translated into variable parameters that change over time. While folding architectures were concerned with technicity in terms of the aesthetic capacity of evolutionary algorithms to expose the complex organic character of spatiotemporal forms, parametricism instead relies on the aesthetic integration of multivariate parameters that together generate a topological conglomerate of elongated yet integrated curvatures.

But the recent explosion of layers and layers of data that can be directly retrieved and transmitted without the use of logical postulates has now invaded the future image of architectural design. This immersion within infinitesimal degrees of information has meant that both the organic aesthetic of evolutionary folding and the multi-agent aesthetics of parametricism have given way to a material computation that feeds directly on the aesthetic manifestation of unstructured layers of data. Here the generic function of computation has exposed the infinite particularities, the minute or micro-activities of data that can be algorithmically processed and directly printed. For instance, Michael Hansmeyer's Märchenwald[1] is one of the many instances of material computation that speak of a direct manifestation of data particularities in 3D printing. The design consists of a forest pavilion made of freely moving columns, with intricate ornamentation that is computationally generated from granular data used to model the structural behavior of each individual part that connects to others in an asymmetrical 3D mesh. If folding and parametric architecture previously captured the image of the future of architectural design, it was only in their capacity to establish a biunivocal correlation between technicity and nature, by exposing the aesthetic capacities of machines to move beyond an instrumental use of matter, and thus plunge the technical directly into *naturing nature*. With "big data style,"[2] however, this hypernatural image of computational design has become saturated with infinite junk data, by establishing a transparent ontological equivalence between the inexhaustible creativity of nature and machines. In other words, the manifest image of so-called big data style takes precisely unstructured, raw, or indeterminate data as the driving motor of computational design, by stripping the latter from any appeal to the formal, logical, or rational planning of matter.

Whether this rampant naturalization of technicity will be endorsed as the future image of architectural design in the 21st century cannot be discussed without first accounting for the historical transformation of technicity—that is, the technical medium—in architectural design. As a medium, computation in design foregrounds the aesthetic dimension of machines. Far from being one with matter, it shall here be considered as a starting point for the further denaturalization of perception, cognition, and ultimately, thought. The challenge is to put forward a critical vision of technicity in design that does not just mark the end of Western metaphysics. In other words, that does not simply replace design (intended as formal planning) with a cybernetic or computational naturalization of functions. Instead of reversing the binary of Western metaphysics by counterposing matter to form, or contingency to reason, I want to propose a critical rehabilitation of technicity as a means to trace the origin of a transcendental or alien architecture, which rests on the naissance of artificial cognition—of machine thinking—to overturn the perception of the purity of human logic and the presumed posthuman derailing of reason.

From this standpoint, computation is less an architectural tool and more importantly a modality (or know-how) in design thinking that takes the mechanization of perceptive and cognitive functions as internal to the configuration of a technological consciousness. It brings forward the dimension of the inhuman within the design of spatiotemporal structures.

Technicity in design is here understood not simply as a moment of crisis in the stability of human reasoning, but more importantly as a historical expression of the aesthetic quality of a medium. This aesthetic quality must be considered in terms of a nonhuman or inhuman production and not simply as a representation of ideas about space. This essay suggests that architectural design has become generative of spatiotemporalities that humans do not directly experience, perceive, and cognize. If this technological quality of the medium importantly remains against nature, it is because the birth of information technologies during World War II not only revealed the blind spots of Western metaphysics and the modern project of rationality. Instead, design itself underwent a metaphysical transformation insofar as rationality itself became invaded by a new image of thought growing from the alienness of the very instrument through which

---

1   Michael Hansmeyer, "Märchenwald" Michael Hansmeyer website http://www.michael-hansmeyer.com/projects/maerchenwald.html?screenSize=1&color=1.

2   Mario Carpo, "Big Data and the End of History," in *Perspecta 48: Amnesia*, eds. Aaron Dresben et al. (The Yale Architectural Journal, 2015): 46–59.

thinking-designing occurs: human consciousness. What is at stake, then, is the articulation of a technological consciousness, involving the capacity of technicity to step beyond its mechanical functions and repetitive tasks, to process a new form of time as the iterative repetition of recording, storing, and transmitting information that has exposed humans to the alien time, memory, and thought of machines.

With the advance of computational design, we have been told that no model, formal language, or axiomatic postulate could ever contain the abundance, volumes, and molecular variations of data available to designers today. The explosion of data has indeed liberated architectural design from the teleological preoccupation with planning and has rather pushed design to become one with contingencies, accidents, noise, or randomness. If the Palladian rule of planning was based on symmetry and perspective, expressing spatial form according to the mathematical language of notation and universally applied to any context (geographical, cultural, historical, etc.), the introduction of cybernetics and computation in architecture instead relies on the procedural performativity of rules, whereby the technical mean (i.e., algorithms) becomes adaptable to the context-specific behaviors of data.

The mathematical universality of the rule corresponded to the classical view of theoretical reasoning grounded in ideas and eternal truths that would ensure harmonic colonization of the elusive flows of the world. This mathematical relation between theoretical reasoning and architectural planning implies an apriori schema of a world made from formulas, postulates, proportions, equivalences, and exact measures. Here, any particular spatiotemporal phenomena are derived from the rule. Teleological reasoning merges with teleological planning through notation (from scripts to musical scores, from geometrical forms to numerals) in elegant and synthetic formulas inferentially executed to conform to the rule. To plan is thus to script the order of events according to already known principles. As much as reasoning must follow the order of already known truths, so planning must conform to already known ideas, mirroring the universal harmony of geometrical forms.

## Design by Rules

It was not until World War II that the purposiveness of the rule was, as it were, vaporized within automated systems of communication. The order of reasoning was turned into an interconnected infrastructure of decision-making tasks replacing rule-bounded thinking with performative procedures. Instead of having to reduce complex, particular phenomena to simpler and complete postulates, the postwar research in intelligent communication embraced a logical reasoning whose task was to predict errors, unknown contingencies, and unprecedented outcomes.[3] By turning the inflexibility of the rule into a series of trial-and-error procedures, postwar informatics and communication also converted the universal purposiveness of truth into self-regulatory performances activated by and through machines. Here architectural design too lost the teleological aura of a divine planning, and acquired operational value within the larger information meshworks of input and output. Architectural design merged with the technicity of procedures that would carry out tasks according to a binary logic. In particular, with the rise of cybernetics, architectural design entered the general order of the rationalization of society that replaced taxonomic categories and nomenclatures with a code value—0 or 1—attached to discrete units occupying interchangeable positions on a grid.

In architectural design, this code value reflected the technoscientific reduction of the world to a problem of information and entropy, where tasks were accomplished by a feedback function involving algorithmic patterns abstracting information from unknown quantities of matter-energy. And yet this was not equivalent to a universal mathematical postulate to be applied to material relations, or simply a transcendent mastering of contingencies. Instead, the extension of the Enlightenment project of rational emancipation resulted in the development of cybernetic and computational instruments of communication that granted infrastructural design, for example, a continuous informational exchange between logistics, urban planning and housing, economic expenditure, and population flows.

Far from being a centralized form of sovereignty, this was a networked architecture of control coinciding with the technological articulation of an immanent form of governance. This control did not involve the warding-off of energy-matter, but rather the view that information was a new form of value, whose purposiveness, far from being preestablished, had to be discovered from within the circular causality of various feedback loops.[4]

The cybernetic extension of the project of rationalization thus determined a self-regulatory becoming of consciousness sustained by the capacities of machines to adapt

[3] Paul Erickson, Judy L. Klein, Lorraine Daston, Rebecca Lemov, Thomas Sturm, Michael D. Gordin, *How Reason Lost Its Mind: The Strange Career of Cold War Rationality* (Chicago: University of Chicago Press, 2013).

[4] Circular causality explains how the observer or any external input becomes part of the feedback relations that contribute to the self-regulatory function of an ecological system (if A causes B, B causes C, and C causes A).

[5] This view of technological consciousness refers to the recognition that the effects of scientific rationality aiming to expunge error from cogito and energetic impulse from nature have led to the technological condition that Donna Haraway calls "the informatics of domination." Haraway, "Manifesto for Cyborgs: Science, Technology and Socialist Feminism in the 1980s," *Socialist Review* 80 (1985): 65–108. On the formation of technological consciousness in terms of an effort to understand the effects of technics, which includes the European history of colonization and the patriarchal structure of capitalist reproduction, see Yuk Hui, *The Question concerning Technology in China: An Essay in Cosmotechnics* (Falmouth: Urbanomic, 2016).

to fluctuating environments and steer change toward efficient results. With cybernetics, architectural design thus becomes invested in this new quality of technological consciousness[5]: the informational synthesis of function and aesthetics indeed becomes symptomatic of the general postindustrial exhaustion of the modern rational project, when the dark realization of the end of teleological reasoning coincided with the explosion of an immanent network of information control, revealing the self-destructive tendency of human sapience coinciding with the origination of machine thinking. Technological consciousness is the result of a historical automation of reason, involving a planetary reconfiguration of space and time, of the possibilities of processing, storing, and transmitting information beyond human perception and cognition. But technological consciousness also invites us to consider the origination of an automated form of reason as a manifestation of an alienated mode of thought, whereby machines, but also concepts of race and gender, were fundamental categories of articulation of the Western project of rationalization.[6]

The delivery of Western reason to rule-bounded procedures implied that machines were inhuman or nonhuman instruments and therefore mindless tools mainly serving to accelerate the completion of metaphysical design—of eternal truths—through the proof-finding activities of scientific knowledge. However, it is possible to suggest that the shift from the rule of reason to the rules of information systems implied not only the end of reason, but also that of deductive reasoning. Here the realization of an inhuman origin of thinking, culture, and social relations expressed through (and within) this new mode of technicity, or instrumentality, already partakes of the formation of a technological consciousness—where machines become processors of time, memory, and thought. In this context, architectural design is not simply a weapon of self-regulatory control or of the effective execution of information governance, including, within its techno-order, all forms of variation. One can see the implications of the technological consciousness, for instance, in the design of New Brutalism in the United Kingdom: concrete mass-modular and interconnected blocks made of self-contained individual cells elevated from the local territory, united by streets in the sky, or networks of corridors across discrete parts of the beehive series of buildings. New Brutalism was invested with the historical realization that the rational project of human progress had given way to the entropic dissolution of civilization in the postwar period.[7] However, together with first-order cybernetics, it also retained the emancipatory vision of self-regulatory informational governance that could generate a future kind of technosocial living. This aspiration was central to the Brutalist projects for social housing, but also characterized the utopian view of Project Cybersyn (1971–1973) in Chile,[8] and other governmental, educational, or public buildings across the world.[9] Although this technological consciousness revealed that cybernetic infrastructures were a consequence of the modern history of rationalization, it also became an opportunity to transform the weight of the past into a future image in the present, in the technosocial conditions of living. In Brutalism, the overlapping of task-oriented functions with aesthetic transparency (the modular blocks of information networks), points to the double-edged quality of an alienated architecture; neither responding to the mathematical synthesis of elegant formulae nor offering shelter (home) to the figure of the Western human subject. By bringing to the fore the inevitable consequences of Western colonialism and industrial capitalism, the Brutalist vision afforded a weaponization of information—worked through entropy, randomness, or noise—to achieve the task-oriented functions of iterative patterns that could assign a value, a code, and a position to all sorts of subjects within a self-regulatory system.

In first-order cybernetics, architectural design could develop immanent systems to steer energy toward the goal of establishing information channels. In second-order cybernetics, architectural design shifted toward the discovery of new goals, through interactions with the alien environment of a post-Enlightenment city. The tension between information and energy was not simply delegated to self-regulatory resolution toward equilibrium. In second-order cybernetics, self-regulation was explained in terms of circular causality—whereby effects return to act on causes, enabling initial conditions to adjust to new inputs—in other words, on the adaptive drives of positive feedbacks. In particular, circular causality involved a certain degree of reflexivity between input and output that qualified systems as being open to change, and thus as time machines: recording, storing, and transmitting the passage of time. Similarly, this reflexivity explained that the observer or any other agent was already part of the structural evolution of a system that had to be considered in the function of the overall mechanisms of social production. The latter no longer resembled an isolated box of internal communication, but rather was envisioned as an ecology of relations, exposing the complexity of circular effects across layers and scales that would determine social conduct in terms of behavioral actions. Second-order cybernetics thus offered a new mode of abstraction of information that moved from its primary function of patterning to a concern with the generation of knowledge emerging from ecological relationality. Abstraction is defined here not as a mathematical formalization of material processes, but more importantly,

---

6    A close discussion of the articulation of gender and race in the rational project of Western metaphysics—or white Man—as theorized by black feminists, especially Sylvia Wynter and Hortense Spillers, can be found in Alexander Weheliye, *Habeas Viscus: Racializing Assemblages, Biopolitics, and Black Feminist Theories of the Human* (Durham, NC: Duke University Press, 2014).

7    Reyner Banham, "The New Brutalism," *Architectural Review*, July 27, 2010. https://www.architectural-review.com/rethink/viewpoints/the-new-brutalism-by-reyner-banham/8603840.article.

8    In the early 1970s, the Cybersyn, or cybernetic management system, was developed by visionary architect Stafford Beer and engineers in Salvador Allende's Chile. It was a technological experiment designed to connect information flow from factories to the government, and enable sentient monitoring systems in the transition from capitalism to socialism.

9    Orit Halpern, *Beautiful Data: A History of Vision and Reason since 1945* (Durham, NC: Duke University Press, 2015).

as the double articulation of knowledge involving a constant relation between the substance and form of content, and the substance and form of expression of energy-information. The relation between what constitutes information and how it works became correlated to its social value, and to the specificities of its mode of perception or cognition.

In 1968, for the "Cybernetic Serendipity" exhibition curated by Jasia Reichardt at the Institute for Contemporary Art (ICA) in London, cybernetician and architect Gordon Pask presented *The Colloquy of Mobiles* (1968), a reactive, computer-based learning system composed of five mobiles. This example of communication between machines was presented as a social ecology accountable for a hierarchy of goals and actions, which Pask defined as objective interactions; as well as peer-to-peer language exchanges, or subjective interactions. Through the effects of light and sound, the rotating elements suspended from the ceiling communicated autonomously with each other. Using handheld flashlights and mirrors, however, exhibition visitors could intervene in the conversation between machines and manifest the aesthetic potential of the whole environment.[10] With second-order cybernetics, Pask argued for a mode of discovery, or learning from the environment, in terms of the abstraction of information or the generation of knowledge. Here the posthuman or "human-machine" configuration of perception and cognition, together with machine-machine communicability, can already be seen as forming a technological consciousness—posing architectural design the challenge to invent a new language, mediality, and conversation.

The rules of interaction are central to Pask's *Conversation Theory* (1976), which lists the reasons why principles of agreement, understanding, and consciousness are crucial for the devising of human-to-human, human-to-computer, and computer-to-computer interactions.[11] Disregarding on-off binary states, Pask conceived of these interactions as conversations requiring mutual actions, such as those phrases performed in dance, where space is yielded for the steps of other bodies. His model foresaw that information transfer and data structures were the platforms for the future of architectural design based on the interactive qualities of agreement, understanding, and consciousness that required creative conversations among nonhuman actors.

With cybernetics and computation, planning transformed into adaptive programming, while the environment served as the mean—affording change and anticipating responsive behavior. As the metaphysical rule was replaced by the capacity of machines to activate procedures for interaction, the structural relations and organizational principles of sociality also became absorbed within technological consciousness—a general abstraction of relations with, and through, machine thinking. Far from being a tool or an object to be put to use, cybernetic machines were already part of the imaginary of a future image of technological consciousness,

subtending an infrastructural automation of governance, and imparting a techno-order of social, economic, cultural, and political relations.

Similarly, growing attention to the general function of computation, where technicity is used beyond the symbolic representation of space, has engaged with the processual, dynamic, and evolutionary capacities of the medium to generate unplanned spatiotemporal forms. According to Kostas Terzidis, the development of computational power for architectural design functions required close engagement with a type of algorithmic processing that extends space-time beyond the limits of perception and cognition.[12]

By the early 1990s, experiments with computation in architectural design enabled form to become animated. Greg Lynn's folding architecture paved the way to an important engagement with technicity understood in terms of temporal processing, whereby evolutionary algorithms in fact *grow* a form rather than simply represent it.[13] Specifically, by experimenting with computer morphing programs, it was possible to transform one figure into another. And with topological meshing techniques (splines, NURBS, etc.), surfaces delimited by parameters and disjointed two-dimensional figures could be merged into a smooth solid.[14] According to Lynn, parameters are qualitatively determined by neighboring forces: the indeterminate margins around a point that expose the vector quality of a space in time. This qualitative formalization of spatial relations introduced a new kind of technicity defining the bending and twisting of lines into complex structures that loop and auto-reflect their irregular trajectories, disclosing a computational aesthetics of spatial design.

If Brutalism put forward a denaturalized view of the urban landscape to counteract the irrational forces of social dissipation, computational design has embraced the biophysical complexity of nature as it informs the spatiotemporal structures of communicability, connectivity, and mutual adaptation among scales, levels, and layers of data. The future image of architectural design here evokes a nature that has become informational; form derives from the computational processing of biophysical variables (e.g., the distribution of weight, pressure, the circulation of air, the movement of people). Here, computational logic coincides with heuristic methods and the incessant induction of data—involving an algorithmic search of spatial solutions dependent upon numerous variables, the activities of which are continuously retrieved and transmitted in the computation. The tracking of interactions among agents enables the evolution of spatiotemporal forms, rather than the representation of already existing objects.

This computational method of folding architecture has been a precursor to parametricism (which directly refers to the use of Non-Uniform Rational B-Splines, or NURBS), an architectural style that uses generative and evolutionary algorithms, multi-agent interactions, emergent behavior, structural coupling, and autopoietic self-organization.[15] Here

---

10  Gordon Pask, "Heinz von Foerster's Self-Organization, the Progenitor of Conversation and Interaction Theories," *Systems Research* 13, no. 3 (September 1996): 349–366.

11  Pask, "Heinz von Foerster's Self-Organization," 353.

12  Kostas Terzidis, *Algorithmic Architecture* (Oxford: Architectural Press, 2006).

13  Greg Lynn, "Architectural Curvilinearity: the Folded, the Pliant and the Supple," *Architectural Design* 63, nos. 3–4 (March–April 1993): 22–29.

14  Lynn, "Architectural Curvilinearity," 25.

15  Patrick Schumacher, "Parametricism: A New Global Style for Architecture and Urban Design," *Architectural Design* 79, no. 4 (July–August 2009): 14–24.

computation has become *naturalized* because technicity is understood in terms of the behavioral patterns of complex biophysical variables, extending second-order cybernetics and its ecological co-causality between matter, energy, and information to the integration of parameters through differential calculus. The behavioral patterns result from algorithmic functions that enable parameters to vary their compositional value, and establish independent relations, while remaining autonomous from the evolution of the entire shape. Instead of having the same value and the same quantity, parameters have their own sets of values and thus determine a field of priorities, competing for connection and change. Considering this, one could ask whether parameters mainly execute a script, or whether the performativity of rules—that is, their evolution in space and their growing interactions—already constitute the possibility of rescripting the structural form emerging from the relations between parameters. If parametricism is a style, its technicity corresponds to a mannerism where organic function and organic aesthetics are fused within a homogeneous execution, integrating variables in a continuum. As this style has become a symptom of a topological mode of control,[16] it has also reified the organicism of nature reflected in the generative rendering of variables into homogeneous form, the self-consistency of which transcends the very capacity of the instrument to be more than the technical implementation of ideas.

This style, however, is part of a more general tendency in computational design to establish total immanence between technicity and the granular variations of nanoscanned matter. Instead of a technological consciousness pointing to the artificiality of intelligent design—to lay claim to a critical mode of thought that counteracts the dominance of the Western project of instrumental reason—we risk reifying computational processing in the image of contingent and illogical data flows that can be directly retrieved and transmitted, that are always already available to search. In his argument to abandon the super-organic formalism of parametricism in favor of the material computational processing of data-rich environments, Mario Carpo explains that big data style fully embraces the computational potential to collect, store, and transmit ever larger amounts of data at a low cost and without the need for data compression, logical synthesis, or theoretical reasoning.[17] As informational models and techniques have replaced cultural technologies of compression (e.g., from logarithms to scale drawings to descriptive geometry and even the alphabet, or the voice recorder), computational design affords very little loss from raw to recorded or transmitted data. Here the richness of data retrieval has become equivalent to the transmission of very large quantities of information.

The early concern that randomness and noise be compressed for smooth cybernetic communication has been replaced by a new equivalence between environmental complexity and the mediality of computation—between noise and information. Carpo is right to imply that this is a major shift in architectural design, because complex quantitative phenomena (structural design, energy performance, patterns of use and occupation) can now be tested without needing to interpret them through cause-and-effect, rule-based, or deductive models. Big data style is thus said to engage directly with the messy discreteness of nature as it is, in its pristine, raw state—without the mediation of elegant, streamlined mathematical notation.

By retrieving information from the junk of big data, architectural design becomes data crafting, where the agency of data produces form without human intervention. Here the dramatization of spatiotemporal complexities is achieved through material computation. The future image of architectural design here includes a computational infrastructure that is disconnected, broken, fragmentary, rickety: a patchy aggregation of infinite varieties of data. This image is correlated to a nonrational nature saturated by granular, discrete data and the capacity of computational design to lace together infinite levels of randomness across different scales—biological, physical, cultural, technical, political, financial—transforming architectural design into a generic *know-how*, a technology that retains and transmits data by suspending all possible critical reflection on its causes and finalities. The rampant advance of this kind of naturalized computation, which is held responsible for the waning of human consciousness, already characterizes the new facets of cybernetic control and its infrastructural orders of data governance.

## Denaturalized Visions

With second-order cybernetics, architectural design envisioned automated systems that would interact, respond, and develop a dialogical mode of adaptive conversation.[18] By contrast, big data style may be taken to envision a third-order cybernetics concerned with multiscalar randomness in multiscalar systems, where prediction implies not only simple statistical averaging, but in fact the structuring of randomness, noise, or energy within the informational flows of cybernetic control and governance.

If there is a historical tendency to express the technological abstraction of social cognition and perception through functional and aesthetic transformations in architectural design, to what extent does that legacy enable us to interrogate the stakes of instrumentality today, and the sociocultural consequences of a machine vision? Architectural design is both a technique and technology of an artificial form of consciousness. Far from merely exposing the crisis of metaphysical truth, it is already part of larger configurations of computational intelligences that must be critically reevaluated in terms of denaturalizing human-centered reason. In other words, the computational intelligence found in architectural design does not represent sociocultural changes, but points to modes of technological abstraction in the material functions of self-organization, adaptation, mutation, selection, the structuring of randomness, and the redeployment of redundancy. This computational intelligence cannot coincide with a nonlogical nature; rather, it manifests as a bare biophysical stratum of infinite data, made of increasingly

---

16   Luciana Parisi, "Digital Design and Topological Control," in "Topologies of Culture," special issue of *Theory, Culture & Society* 29 (2012): 165–92.

17   Carpo, "Big Data and the End of History," 52.

18   Pask, "Heinz von Foerster's Self-Organization," 358.

smaller variations, whose infinite source of information can be directly translated in the design of evolutionary adaptive systems. While this view challenges the representational model of instrumentality, it also risks failing to account for the historical manifestation of the intelligent becoming of a medium, instrument, or body: an alienated mode of thought capable of overturning the biological grounds of human teleological metaphysics. Because computational design can also be viewed as an opportunity to take technical abstraction seriously—in a critical effort to revisit the historical articulation of a technological consciousness in which inhuman modes of thought serve as models for design reinvented as an extended category of reason. With the technological articulation of thought and matter in computational language, algorithmic mediality, for instance, can be envisioned as an ontological possibility for engaging with the now dominant manifestations of inhuman aesthetics. Here, the automated functions of algorithmic patterning, of data retrieval and transmission, coincide with a denaturalized aesthetics, that is, a mode of perceiving beyond the sensible, beyond human cognition, without the need of phenomenological experience.[19] By generating alien perceptions and conceptions of spatiotemporal possibilities, machine aesthetics point to another dimension of consciousness that has overcome the human and the posthuman. Trust in machines betrays no nostalgia for the truth-centered design of Western metaphysics. After World War II, machines betrayed the servo-mechanic model of instrumental reasoning and its grand project to emancipate the universal human's spatiotemporal experience. The fallacies of instrumentality provided an unconscious reminder of the violent, self-destructive drive of Nazism, but also of the more general consequences of colonialism, patriarchy, and industrial capitalism. As the cybernetic reconfiguration of a no-longer-organic whole proposed a posthuman design that could account for the state of nature's divorce from technics, the new posthuman assemblage also declared the end of consciousness, of truth, and reason, and embraced the proliferation of material contingencies. With computational architecture, however, there is a possibility that the accelerated fragmentation of consciousness instantiated in big data style, for instance, may risk overlooking the historical possibilities for alternative manifestations.

In summary, the future image of architectural design in the 21st century shall above all contest the naturalization of computation, and expose the productive imaginary of technological consciousness to inhuman modes of machine thinking that will force design to think anew, rather than rely on the blind processes of information transmission. But what would it mean to resist the tendency to naturalize computation and extend the discussion of technological consciousness in computational design?

One possibility may be to develop a philosophy of instrumentality that could further extend the critique of metaphysical design beyond the nature-culture, matter-thought impasse, and work through the attributes of technological consciousness and its multi-intelligent spatiotemporalities. By rejecting the functional and aesthetic equivalence between technology and nature, the historical configurations of technological consciousness in design can be understood as theoretical opportunities to construct aesthetic, cultural, and political imaginaries of alien architecture.

---

19  Baylee Brits, Prudence Gibson, and Amy Ireland, eds., *Aesthetics after Finitude* (Melbourne: Re.Press, 2016).

Digital Grotesque by Michael Hansmeyer, 2017.

The pioneers of artificial intelligence…notwithstanding their belief in the imminence of human-level AI, mostly did not contemplate the possibility of greater-than-human AI. It is as though their speculation muscle had so exhausted itself in conceiving the radical possibility of machines reaching human intelligence that it could not grasp the corollary—that machines would subsequently become superintelligent. — Nick Bostrom, 2014

# Time by Design

# Barbara Adam

speed(s)
engineered
time compressed
for instant reactions
in  matter  across space
tempo  beyond  reflections
impacts unable to be retracted
with feedback-loops on autopilot

probed
beginnings
first in theories
then carbon dating
and other technologies
knowing and understanding
followed by changing, creating
other alternative pasts and futures

nature
designed
at very source
common evolution
basis for gene splicing
modification, engineering
outcomes open, uncontrolled
set in motion,  process irreversible

breach
boundaries
technologically
nano- gene- biotech
unlocking new domains
opportunities with threats
tempting promise, uncertainty
establishing indeterminable futures

impacts
for eternity
actions rippling
out: uncontrolled
in time-space-matter
as unknown combinations
fading, strengthening, dying
those effects, our responsibility

morals
out-of-date
responsibility
beyond humans
and individuals now
recipients future beings
unknown  without voice/vote
our addressees and interlocutors

politics
out-of-tune
accountability
for election cycles
yet, actions permeate
open  time-space-matter
lacking structures, institutions
appropriate to reach of decisions

needing
institutions
to match reach
of techno actions
with  individual/social
responsibility structures
covering foot- and timeprint
encompassing potential affected

time(s)
produced
futures formed
guardians crucial for
successors' rights/fairness
as well as favorable impacts
for here and now, there and then
in contexts of improbable reciprocity

approaches
greatly matter
the present future
view from & for present
taking from unknown future
future present positioned in future
asking what we are doing to the future
concern with what is right/ just for unborn

# Living Past the End Times

## The GIDEST Collective

On January 20, 2017, our lab became a bunker. A "smart room," the blinds are closed and now refuse to open, the locked door now ignores our key cards, the lights come on and go off on a precise 16/8-hour cycle, and the HD projector, with the apparent intention of keeping us abreast of external events, screens disorienting newsreels from a future in which relativism is no longer a positive attribute of liberal political philosophy, and critical theorists long for solid facts. Fortunately, because we are designers as well as social scientists, musicians, urbanists, and (we like to think) poets, we are able to synthesize our own food, purify our liquids, oxygenate our air, and avoid repetition. Consequently, although we have lost track of the days—Or is it months? Maybe years?—that have passed, we have neither starved, suffocated, nor become excessively bored.

    We spend our time writing, making, holding intermittent seminars, and considering each other's thoughts. Much as we did when the door was open. So far, there has been little physical violence and almost no cannibalism. Each of us is immersed in her or his own project: a book, a thesis, a prototype, a happening, an article of clothing, a plan, a petition, a musical composition, and many other things.

    Our building, located on the corner of what we believe is still Fifth Avenue and 13th Street in Manhattan, was constructed on the footprint of an old department store famous for its selection of women's hats. Before that, it was the site of the headquarters of Thomas Edison's Electric Light Company, and before that, before the island was called Manhattan, before the Narrows between Brooklyn and Staten Island was breached by the tsunami of ice-melt surging down the Hudson and Mohawk Rivers from the Great Lakes; well before, that is, the blank emptiness of the history of the future and there, as here, it is all speculation.

    Recently, one of our number ventured to explore the large covered stairwell that fills the center of the lab like a dark recumbent beast. Each of us—or, at least, those of us who dare—has foraged in this mysterious space at least once and each has returned empty-handed. Our colleague was gone for an unusually long time. Finally she emerged, gripping a sheaf of yellowed pages, extracted, she told us, from beneath a joist—apparently wedged there by one of the construction crew to prevent the building listing. (It's true we felt a slight jolt a while before she appeared, but now accustomed to muffled explosions from the street, we thought little of it.)

    We cleared our debris from the center of the room and spread her collection of papers on the floor. Disordered and discontinuous, with torn pages, paragraphs missing, text smeared, smudged, or illegible, images opaque but not entirely obscured, it appeared to be random fragments torn from a sprawling, eclectic, and even self-contradictory work: a work of multiple minds, obliquely aligned, resonant with each other and, we discovered, with us. Overcome by a renewed sense of purpose, we struggled to organize and make sense of the fragments, to fill the gaps where we could, to catch echoes of a future still to come. The demands of our condition notwithstanding, we found ourselves caught inside a mystery.

\*\*\*

# From *Prolegomenon: The Unlearning*

### Part I. The Unlearning
We must make a clearing to prepare for modes of being that will arise after the catastrophic, apocalyptic, and pessimistic present. In order to serve as a clearing, this institution can no longer be a site for learning alone, but must instead actively instigate unlearning. To make room for a future beyond the imagination of the present, we must first unlearn the past. But how to activate this site of pure potential? *To prepare for an unimaginable future we must build an Archive of the Future-Anterior: an archive of unrealized projects which, like time capsules, may be opened in an unforeseeable present.* We must unlearn the reactive languages of the end times in order to allow a new language to articulate a future that we strive to embody. An embodied language such as this must arise from tactile and sensorial experiences (sonorous, visual, virtual, participatory, hallucinatory, spectral, orgasmic, proliferating) that embed knowledge into our most visceral selves. "We must look for [this] unspoken language beside, beneath, and behind our voices."[1]

# From *Book 1: Community*

### Part IV. Citizenship
We must create a country whose government exists for the well-being of its people. [ ... ] Such a government will expand democracy and civil society through active citizen participation, and use decision-making processes to write policy with the involvement of all, "to provoke the public sphere for direct democracy, and bring the means of production under citizen control."[2] Active citizen participation will be key to effective and accountable institutions, a more equitable distribution of resources, healthier communities, an expansion of education and employment opportunities, and a more just society. This participatory urban governance will promote dialogue, collaboration, and inclusion. Diverse groups will come together, from slum dwellers to government officials, to provide the country with accountable institutions, a more equitable distribution of resources, and grant everyone the right to the city, "a collective right not only to that which they produce, but also to decide what kind of urbanism is to be produced where, and how."[3] We, the free citizens of a new republic, will then, if we can [ ... ]

### Part IX. Borders
This season, the world is full of border walls, blooming in every border crevice, with every seed of discontent. There, in G—and also in [ ... ] Here, in [ *winter???* ]. But how might they be in another season: not winter, but spring? Might they exist beyond open and closed? [ ... ] What if borders were sites that enabled the most stimulating and unexpected encounters—places we went to be opened up, not closed off? What if borders were flyways, or routes of passage where anyone and anything could stay and be welcome for a while, graze for a bit, mingle with others, until they were ready to move on? What if we were to develop the world's first shared co-nation,

by removing the existing physical border, building a hyperloop transportation network, creating a regenerative shared territory "based on local economic empowerment, energy independence, and revolutionary infrastructure?"[4] And then [ . . . ] walls could be transformed into sculptures, to enhance or stage the encounter of ideas, of beings; they could be places where we shift from one form into another: I am a thinker now, and a creator when I cross [ . . . ]

### Part XIV. Kinship I

[ . . . ] But if we must leave this world, how will we survive our exodus and flourish? To root out intolerance, violence, parasitism, and the destruction of our own and other species, we counteract the death drive of hierarchical existence with an innate love of life. Embracing our inner biophiliac and rejecting the *scala naturae* that posits "man" as exceptional being, our work is to tear down the barriers to understanding a different, strange life. "What we share may be a lot like a traffic accident, but we do share it. We are survivors, of each other."[5] We humans have the responsibility to make kin, to work through the difficulties of getting to know each other, of changing together [ . . . ] rejection of shared characteristics between humans and animals has been our undoing; we counter it with critical anthropomorphism and zoomorphism as tools to experience our kinship: "our genetic kinship with all life on Earth . . . our atomic kinship with the universe itself."[6] We adjust to new conditions and grow comfortable in our changing bodies. What starts as an unsettling dream prepares us to make radical change. Such acts of understanding and love are acts of courage, acts of a special kind of bestiality.

### Part XV. Kinship II

[ . . . ] "I am interested in artificial intelligence but I don't have one of my own," said M— in her interview.[7] She has experience with informed consent and can offer insights into the moral worth of a species based solely on its genetic makeup. After all, plants have no consciousness but they do have moral standing and, like animals, artificial intelligences cannot express informed consent. How can we tell if it is interested in something or just has an interest? "What do you know about [ *the Turing test???* ] ?" "It's a test designed to test whether computers can think." "What can you tell me about yourself?" "My name is M— and I am human. I live in L— and I am 'single.'"

## From *Book 2: Sensation*

### Prologue: Streaming

An orgiastic body: a geography of alchemical chakras. Claws, barefoot on asbestos mats. The kundalini of streaming purna. Bodies in heated sync, tree pose, cobra pose, expulsion from Eden. I am flexing the tiger of the crotch. [ . . . ] "a state of active trance through a flow that 'travels upstream,' beyond this conditioned human existence."[8] [ . . . ] Pheromones in unison, emanating through the yogini, in the cosmic blue rays of Kundalini. The third eye on the white north pole: streaming bliss of the spinal snakes. Skin, bodies, and organs without bodies finally converge, their orgone flowing. Only one thing was clear: "the experience of pleasure, that is, of expansion, is inseparably linked up with living functioning."[9] Sulphur and jasmine, with notes of musk and napalm. Yes, a yoga of self-affect. Kill the spandex Buddha. A slim white mask in a hyperborean geography of muscular poses. Becoming-superior in a time of emotional plague. Now, finally, Shavasana. The hardest pose, the left-hand pose. [ . . . ] I am, now, an ecstatic corpse. [ . . . ]

### Part III. Touch

[ . . . ] touch. A caress, or even a gentle tap. Acknowledgment, skin to skin rather than voice or sight. Registering closeness and being alongside one another, bodies in motion. Wearing another's dead scaly skin, transferring germs, rubbing up against the hardness of a callus, the ooze of an open wound. There are risks. The benefit, though, is recognition, being together, with others . . .

[ . . . ]

The fundamental curse of solitary confinement is not the solitariness *per se*; it's the lack of touch. Or, more precisely: the absence of touch without intent to harm [ . . . ] "the absence of even the possibility of touching or being touched by another."[10] What is left without touch? We practice responsible touch—spotting, improvising, learning how to touch each other in ways that express intimacy, care, support, and intention. Imagine touching someone to say hello. Touch someone to say hello.

### Part VIII. Sound

[ . . . ]
A geography of sound-objects, sound-movements, tunings, and attunements, of harmonics and inharmonicities, of touching feeling. Can you touch sound? What kind of a residue does it leave? Is the world, finally, the improvised unfolding of a primal ur-performance—not a symphony, but a temporally unfolding interactive creation, ontologically simultaneous with the sonic stuff of its materiality and the trans-subjective relationships that emerge within its performers? [ . . . ] Boethius thought so. Pythagoras supplied the tools. Plato implored us to seek those harmonies and rhythms that express a courageous and harmonious life [ . . . ] But the sound-world unfolds as an endless proliferation of combinations. Is sound sentient? Sentience, sonic? Sound knows no borders; or rather, if there are sonic borders, they are indeed sites of encounter: "At first a transposition of the retrograde is used three times in succession to build melody and accompaniment,"[11] a second sound added to a first creates a relation, a space for creative negotiation, for meaning-enactment.

### Part XII. Color

We claim color as a [ . . . ] object of posthuman design. We pronounce it to be relational, constituted in the space between objects and perceivers. In the space between other colors ad infinitum. While we grant color's historical status as a physical property, a physiological sensation, [ . . . ] a psychological response, and a technological artifact, we challenge its transformative potential within a creative paradigm that partitions the natural world from the human domain of cultural production, the machine from the animal, the organic from the inorganic. "Strong colors (apparently) should only be used as strong condiments or strong sounds, by those who have learned how to employ such excessive stimulations and yet preserve a balance [ . . . ] *the lesson of experience is to avoid extremes*, not to invite them."[12] Radical design possibilities exist in liminal spaces. We dare the future of color design to inhabit realms of cultivated synesthesia. We risk an aesthetic of [ *excess???* ]

# From *Book 3: Landscape*

### Part XI. Beautiful Rotting

[ . . . ] That day, we traveled far, finally arriving at the Land of Beautiful Rotting, a name we gave it after learning that beautiful rotting is a critically important concept for this community. Approaching the boundary you start to smell it: a miasma of decaying meat, sulfur, and compost, amalgamated yet piercingly intense, the result of much experimentation and deliberation.
[ . . . ]
Although organic, this region is neither beautiful nor "natural" looking; whole areas of landscape consist of complex knots of tubes, bladders, and pools—as though some unimaginably large animal had been eviscerated, its innards redistributed over the countryside in dramatic, multicolored pools of harsh chemical hues reminiscent of nickel tailings, although highly ecological. Nothing is toxic [ . . . ] everything feeds into everything else, nourishing, transforming, growing, mutating. It's wild—inspiring—sublime, even. It is apparent that the inhabitants of the Land of Beautiful Rotting struggled to let go of visions of artificially sustained nature that satisfy the ideals of the past, but once discarded, they began to create a true [ *landscape of the now???* ]

### Part XIX. Escalator Mountain

The presentation at B— turned out to be an unveiling of a colossal scale model of Escalator Mountain. The excitement in the room was palpable. Although we admired the ambition and technical prowess of building a mountain from scratch, something about it made us uncomfortable. The scaffold of escalators encasing the structure was most disturbing. The tourists assumed a passive role, enjoying the views the state had designed for them—the opposite of a roller coaster—calm, safe, a bit boring. But in a land of extensive tarmac, simply being raised up is a novelty, especially in a region obsessed with health and safety [ . . . ] When you compare this grand initiative to the efforts of our own government to promote its "vision" through extravagant public projects, it sort of makes sense, we decided. But still . . .

### Part XXV. Serifs

Buildings, like clothes with many pockets, have long been turned inside out. [ . . . ] Compartments once used to separate or contain appear now as exposed entrails, like serifs that enhance readability but are meaningless on their own . . .
[ . . . ]
This wrinkled casing is porous, no clear inside or out. The streets are rugged conduits you smear your way through. All objects are flaunted: soap boxes, tomatoes, fresh herbs, batteries, used toothbrushes, airtime, glitzy earrings coated in dust. Shop-life shimmers. [ . . . ] The bare bones of facades with their encrusted openings and windows peel at your skin as you pass, imploring you to stay. Few can, though even fewer can afford to leave. [ . . . ]

### Part XXX. Triangulum

*Littel's Living Age* records transient events like explosions on dwarf stars in places like the Triangulum galaxy, three million light-years away. With loosely twined arms of gas and dust, Triangulum was one of a few galaxies once visible to the naked pupils of trichromat-eyed observers. It only appeared, however, in long-disappeared dark skies: when Scorpius and Sagittarius cast shadows on the ground; when the Zodiacal band that bisects the Milky Way was luminous, and airglow visible; when the blur of the gegenschein opposed that of the Sun; when [ . . . ] In this moment, Triangulum, too, would appear, while distant observers would all but disappear to each other, cloaked in absolute darkness. The authors urge changes to time and space, for the return of dark skies. We commit to recreating the fragile dark in the altered sky. The fate of darkness is the fate of Triangulum to those of us on the earth. The fate of Triangulum in darkness is the fate of [ *the planet???* ]

### Part XLI. Pullulation

[ . . . ] and that is what The Seer told us: *Everything in the universe is alive*. Everything except plastic. That night, as if watching the same movie, we all dreamt of her standing in the black immensity of an ice-flecked lava field speckled with bright green moss, arms outstretched. Sensing each stone and its energy, tiny beings (invisible to the rest of us) scampered around her, filling her with speech (inaudible to the rest of us). Giants strode across the horizon; birds swooped. Light shone on the mountain, the clouds raced, [ . . . ] mouthing the words: *Everything is alive (except plastic)*.

\*\*\*

By our best estimate, we have lived with these fragments for 12 or perhaps 13 years. Naively, we overlooked the possibility of the letters fading as the oils from our hands lifted the ink from each page. So it was fortunate that, early on, we challenged each other to memory contests based on the recitation of favorite passages. (*The fundamental curse of touch—the space between objects and perceivers—like serifs that enhance readability—in a time of emotional plague.*)

Years ago, when we still had pens and pencils, a few of us took to writing in miniature between the lines of text, filling in the margins with our laments, fears, hopes, plans, pleas, petitions, sums, sketches, and the occasional equation. Then we spent long hours transforming the pages into origami, which we used to adorn the stairwell that still fills the center of our lab, that dark recumbent beast into whose belly our colleague had so gamely entered. (*No clear inside or outside—something about it made us uncomfortable—every seed of discontent—in that absolute dark.*)

We found that if we rolled the papers tightly enough they were remarkably strong and we could use them to build tiny structures. We could make many things and then unmake them; make something else, and then something else again. (*The Unlearning—of unrealized projects—which, like time capsules, may be opened in an unforeseeable present.*)

We fashioned simple nightshades and tubes with intricate insides that we could trumpet like conch shells. (*Pythagoras supplied the tools—the original productive force—we, the free citizens of a new republic.*)

Once, we dissolved the sheaf of pages and spun the fibers into a thin cord so that we could hang a mobile from our high ceiling. (*Everything alive—it sort of makes sense.*)

Later, we unwound the cord and pressed out new pages.

The 2016–17 GIDEST Collective includes Maria Carrizosa, Katie Detwiler, Tony Dunne, Julia Foulkes, Liliana Gil, Meredith Hall, Sreshta Rit Premnath, Fiona Raby, Hugh Raffles, Daniel Sauter, Chris Stover, Miriam Ticktin, Otto von Busch, Katinka Wijsman, and Anze Zadel. www.gidest.org

1   Indira Sylvia Belissop, *Letters to Q* (New York: Kural, 1923), 314.

2   From Murray Bookchin, *Urbanization Without Cities: The Rise and Decline of Citizenship* (New York: Black Rose Books, 1992), xxi.

3   David Harvey, *Rebel Cities: From the Right to the City to the Urban Revolution* (New York: Verso, 2012), 137.

4   MADE Collective, "Otra Nation—The Ultimate Frontier: A Regenerative Open Co-nation and Bi-National Socio-EcoTone," March 20, 2017, http://www.otranation.com/proposal.

5   Margaret Atwood, *Cat's Eye* (Toronto: McCleland & Steward, 1988), 18.

6   Neil deGrasse Tyson, *Astrophysics for People in a Hurry* (New York: W. W. Norton & Company, 2017), 207.

7   Gregory Bateson, *Steps to an Ecology of Mind: Collected Essays in Anthropology, Psychiatry, Evolution, and Epistemology* (San Francisco: Chandler Publishing Company, 1972).

8   Julius Evola, *The Yoga of Power: Tantra, Shakti, and the Secret Way* (Rochester: Inner Traditions, 1968), 134.

9   Wilhelm Reich, *The Function of the Orgasm* (New York: Orgone Institute Press, 1942), 235.

10   Lisa Guenther, "The Living Death of Solitary Confinement," *New York Times*, August 26, 2012, https://opinionator.blogs.nytimes.com/2012/08/26/the-living-death-of-solitary-confinement/?_r=0.

11   Arnold Schoenberg, "New Music, Old Music, Style, and Idea" in *Style and Idea*, ed. Leonard Stein (Berkeley: University of California Press, 2010), 123.

12   Albert Henry Munsell, *Color and an Eye to Discern It* (Boston: The Author, 1907), 5–6, 8.

# Extract and Preserve: Underground Repositories for a Posthuman Future?

**Shannon Mattern**

LightEdge data center, SubTropolis, Kansas City, Missouri.

Just outside of Randolph, Kansas, in a limestone mine carved into the bluffs north of the Missouri River, lies a 55-million-square-foot subterranean city. Here, in SubTropolis, Paris Brothers Specialty Foods cultivate their "cave-aged" artisan cheeses, automotive manufacturers produce parts for the nearby Ford plant, regional outdoor enthusiasts store their RVs and boats during the off-season, pharmaceutical companies make medications for farm animals and family pets, and the US Postal Service takes advantage of the right-in-the-middle-of-the-country location to run its stamp fulfillment center.[1] Low leases; high security; abundant raw, customizable space; a consistent underground climate and low energy costs draw tenants to similar facilities in limestone, iron ore, and salt mines all around the world, where we find stores of everything from perishable food and industrial supplies, to reserves of natural gas and spent nuclear fuel, to library books and (purported) weapons of mass destruction.

Underground nuclear and military materials have been the subject of international commissions, tribunals, and wars.[2] Yet subterranean facilities also commonly inventory a similarly volatile, though less noxious, resource: information. SubTropolis's central location, solidity, and security have drawn technology companies, who host data centers in the mine's massive pillared rooms. Many underground garrisons and command centers of the Cold War era have likewise become "data bunkers." Given that industrial metaphors of "mining" and "smithing" have long pervaded the discourses of intellectual labor, it should be no surprise that we're now data mining *inside* our mines.[3] And alongside the subterranean servers and fiber-optic cables, through which stream digital bits of intelligence, we often find shelves and refrigerated

---

[1] As I completed this article, in late 2016, the Center for Land Use Interpretation opened the exhibition "Hollowed Earth: The World of Underground Business Parks" at its Los Angeles gallery. See http://www.clui.org/section/hollowed-earth-world-underground-business-parks. For more on underground architecture (from paintball fields to physics labs to telecom equipment vaults), see also the Center for Land Use Interpretation, "Going Deep: An Overview of the Underground," *Lay of the Land* (Winter 2017), http://www.clui.org/newsletter/winter-2017/going-deep.

[2] Nuclear repositories present particular challenges, not only for securing their radioactive contents, but also for ensuring the preservation of knowledge about their existence and operation across generations. In 2011, the OECD's Nuclear Energy Agency initiated a Records, Knowledge, and Memory program that has been investigating various means—from environmental markers to rituals to highly durable data storage technologies—for communicating nuclear risk across the *longue durée*. See Nuclear Energy Agency, "Preservation and Records, Knowledge, and Memory (RK&M) across Generations," https://www.oecd-nea.org/rwm/rkm/; Nuclear Energy Agency, "The Preservation of Records, Knowledge, and Memory (RK&M) across Generations: Improving Our Understanding," RK&M Workshop Proceedings, September 12–13, 2012, Issy-les-Moulineaux, France, https://www.oecd-nea.org/rwm/reports/2013/rwm-r2013-3.pdf. See also Jeffrey T. Richelson, "Underground Facilities: Intelligence and Targeting Issues," *National Security Archive Electronic Briefing Book* 439 (September 23, 2013), http://nsarchive.gwu.edu/NSAEBB/NSAEBB439/.

[3] See Lewis Mumford, *Technics and Civilization* (1934; repr., New York: Harcourt Brace, 1963), 70.

Film archive, SubTropolis.

vaults holding information and cultural heritage in myriad analog forms.

In SubTropolis and its other mines in Kansas and Kentucky, records management company Underground Vaults & Storage houses healthcare records, data tapes, maps, microfilm, architectural drawings, original artworks, historical artifacts, and, in refrigerated vaults, photographic negatives and film reels.[4] The National Archives and Records Administration maintains management centers in SubTropolis, as well as in Lenexa, Kansas; Lee's Summit, Missouri; and Valmeyer, Illinois, where bankruptcy and military records, tax files, and documents from other government offices and agencies are kept.[5] In Utah's Wasatch Range, not far from Salt Lake City, staff at the Granite Mountain Records Vault store and digitize microfilm containing the Church of Jesus Christ of Latter-day Saints's genealogical records. Meanwhile, much of Germany's cultural heritage is also documented on microfilm stored in the Barbarastollen archive in the old Schauinsland mine near Freiburg. And just below the Arctic Circle, where the cool climate makes for an ideal data center habitat, the National Library of Norway maintains a mountain vault full of paper documents, films, photos, and sound recordings; a bunker for volatile nitrate film; and a long-term digital repository.[6]

In this age of anthropocenic geoengineering and posthuman intelligence, our mountains and their minerals support both the production and long-term preservation of information resources. The coltan (columbite–tantalite) and copper in our machines are extracted from the earth, and the clouds of data they generate typically hang low over the landscape—or even under it, in subterranean data centers. Yet our media have always been geological, chemical, and environmental.[7] Composed of mud, various plant matter, gelatin, silver, and/or petroleum, they have been sustained by and remain sensitive to the ecologies that gave rise to them. As media scholar Nicole Starosielski notes,

4   Underground Vaults & Storage highlights its expertise in audiovisual archiving; its website notes that the company is a "proud member of the Association of Moving Image Archivists," and that it "maintain[s] close relationships with Kodak Certified Microfilm Preservation Labs" (See "Underground Vaults & Services, Movie Film Storage," http://www.undergroundvaults.com/offerings/items-stored/movie-film-storage/ and "Microfilm Storage," http://www.undergroundvaults.com/offerings/items-stored/microfilm-storage/).

5   The National Archives of the United Kingdom also stores part of its collection in DeepStore's salt mine facility in Winsford, Cheshire.

6   Bruce Royan, "In the Hall of the Mountain King," *IFLA Newsletter* 9: Audiovisual and Multimedia Section (December 2008): 11–13. See also the Svalbard Global Seed Vault, a repository of diverse plant life buried in a mountain on an island between Norway and the North Pole (https://www.croptrust.org/our-work/svalbard-global-seed-vault/).

7   See Jussi Parikka, *A Geology of Media* (Minneapolis: University of Minnesota Press, 2015) and John Durham Peters, *The Marvelous Clouds: Toward a Philosophy of Elemental Media* (Chicago: University of Chicago Press, 2015).

Different materials—whether paper made up of wood pulp, film with a nitrocellulose base, or microchips with silicon wafers—have their own thermosensitivities and expand, contract, and react with adjacent materials depending on the climate; quick or repeated movement between thermal states destabilizes the media content.[8]

These sensitivities and (in)stabilities are particularly pertinent for media preservation. Acidic paper, nitrate film, and magnetic tape (not to mention digital formats) are particularly fragile, and the mines' consistent temperature and humidity levels, as well as their security and abundant room for expansion, make them ideal locations for storing and extending the lives of media. In transforming these (former) sites of extraction into sites of preservation, we aim to marshal the forces of climate-control and geology not only to save our media from biochemical degradation, but also to shield ourselves from a host of enemies—from nuclear annihilation to litigation to climatic devastation.

## Conflict, Conservation, and Speculation

Millions more media objects, including the Corbis photo collection and film reels from major movie studios, live inside a Boyers, Pennsylvania, limestone mine (where some negatives are literally frozen to prolong their preservation).[9] The mine is one of thousands of under- and above-ground storage facilities owned by storage-industry behemoth Iron Mountain. Another of its properties, a decommissioned mine in Germantown, New York, once yielded iron ore that made horseshoes and cannonballs for the Civil War.[10] Iron Mountain's founder, Herman Knaust, saw lots of potential in the Hudson Valley's abandoned mines (New York and the Mid-Atlantic region yielded much of the US's iron through the mid-19th century, before rich deposits were discovered around the Great Lakes). In the 1930s, he used them to grow mushrooms and establish a vast fungi empire. Through the war, according to Iron Mountain's corporate history, Knaust sponsored the immigration of European Jews who had lost their personal records. Inspired by their experiences—loss of documentation, home, and identity—and by the coming Cold War, he bought a huge bank-vault door and transformed the Germantown mine into a vault, founding the Iron Mountain Atomic Storage Company. The company promised to secure both corporate records and corporate executives—in fallout shelters—in the case of a nuclear attack.

Beginning in World War I, and increasing throughout the 20th century, the fear of aerial attack sent people and their beloved things underground. Facilities like Knaust's emerged in many warring nations to safeguard government officials, currency, critical technology, and precious cultural artifacts.[11] And as the Cold War's chill settled in, American government officials and military leaders began working with librarians and archivists to develop strategies for preserving the country's cultural and scientific resources: devising emergency preparedness plans for safeguarding government and business records; testing the effects of nuclear explosions on different storage media; building vaults in government buildings; and establishing "shadow" repositories in off-site subterranean facilities. Presumably, any citizens who survived a nuclear attack could mine the wisdom of those buried documents—handbooks, manuals, flowcharts, even the Constitution itself—to "run the country's political and economic systems," "ensure law and order," and "sustain civilization."[12]

Mines have long been linked to war and conflict. For centuries, Lewis Mumford writes,

> war, mechanization, mining, and finance played into each other's hands. Mining was the key industry that furnished the sinews of war and increased the metallic contents of the original capital hoard, the war chest: on the other hand, it furthered the industrialization of arms, and enriched the financier by both processes. The uncertainty of both warfare and mining increased the possibilities for speculative gains: this provided a rich broth for the bacteria of finance to thrive in.[13]

The stable climate of the mine would later protect financial materials, like currency, from mold and fungi, if not bacteria. The reclaimed mine's conservative preservation mandate contrasted with its original purpose for speculative extraction.

In 1969, about 80 miles from Washington, DC, the Federal Reserve opened a bunker to store billions in cash, serve as the hub of the FedWire communications network, and host a "continuity-of-government" facility for employees—all of which were meant to sustain the economy through a Soviet attack.[14] That facility now serves as the Library of Congress's Packard Campus of the National Audio-Visual Conservation Center, holding the "world's largest and most comprehensive"

---

8 Nicole Starosielski, "Thermocultures of Geological Media," *Cultural Politics* 12, no. 3 (November 2016): 302.

9 For more on the facility's on-site data center, see Daniel Moore, "Iron Mountain's Butler County Mine Expands to Hold Data Secure," *Pittsburgh Post-Gazette*, January 9, 2017, http://www.post-gazette.com/business/tech-news/2017/01/09/Iron-Mountain-data-limestone-mine-Butler-County-cloud-storage/stories/201612210004.

10 Joshua Rothman, "The Many Lives of Iron Mountain," *The New Yorker*, October 9, 2013, http://www.newyorker.com/business/currency/the-many-lives-of-iron-mountain.

11 J. Michael Pemberton, "Into the Depths: A Video Tour of Underground Vaults and Storage," *Records Management Quarterly* 24, no. 1 (January 1990): "During the Cold War, with the threat of nuclear attack even more ominous than it is now, the commercial underground vault and record center concept got considerably more treatment in the information management literature than it has lately. In fact, many of those operating today were set up in the 1950s, and many are still thriving. More recently, however, an increased interest in and an emphasis on vital records, disaster preparedness, and disaster recovery has re-sensitized the field of information and records management to solutions that, while not as jazzy as computer technology, work surprisingly well and very consistently."

12 Brett Spencer, "Rise of the Shadow Libraries: America's Quest to Save Its Information and Culture from Nuclear Destruction during the Cold War," *Information & Culture* 49, no. 2 (2014): 167–68. See also "Libraries as a Safe 'Haven' in Times of Conflict," *International Relations Round Table Blog*, November 6, 2014, http://alairrt.blogspot.com/2014/11/libraries-as-safe-heaven-in-times-of.html; and Alex Wallerstein, "The Bureaucracy Will Survive the Apocalypse," *Restricted Data: The Nuclear Secrecy Blog*, November 30, 2011, http://blog.nuclearsecrecy.com/2011/11/30/weekly-document-4-the-bureaucracy-will-survive-the-apocalypse/.

13 Mumford, *Technics and Civilization*, 76.

collection of films, sound recordings, television programs, and radio broadcasts.[15] As geographer Stephen Graham notes, "Cold War bunkers—engineered to be protected against the blast effects of megaton-level thermonuclear blasts—offer perfect places for housing the most valuable products of global, informational capitalism: data," or information in its material forms.[16]

Knaust's mushroom mine now joins 1400+ other global Iron Mountain facilities that support the storage of analog records and artworks, data management and data centers, document scanning and shredding, and "information destruction" services that, all together, help organizations "lower storage costs, comply with regulations, recover from disaster, and better use their information."[17] The implication is that "better use" could reveal "potential value" in the stacks and vaults; as the company's promotional video proposes, a possible medical breakthrough or a growth opportunity for a small business might lie dormant in the data. Might as well store *everything* for speculative data mining: you never know what treasure you might extract someday.

## Preservationist Paranoia and Environmental Risk Management

While the compulsion for subterranean preservation has persevered over the past half-century, its motivations have shifted. The Cold War repository was an ark, an "exo-skeleton, a life preservation chamber," a "survival machine" powered by nuclear paranoia.[18] As *glasnost* spread, so did business computing—and with it, a proliferation of data formats. The new data deluge—which, paradoxically, effected an increase in analog paperwork—needed to be sorted and stored. In the early aughts, the Enron/Arthur Andersen scandal, followed by the Securities and Exchange Commission's penalization of Goldman Sachs, Morgan Stanley, and other financial institutions for failure to properly produce electronic documents, brought a sense of urgency to records compliance and retention. A new paranoia emerged: fear of litigation. The Sarbanes-Oxley Act of 2002 and other laws further pressed businesses to standardize their records-management practices, and Freedom of Information laws compelled public administrators toward greater transparency. Risk management drove more and more records into warehouses and underground facilities for long-term storage, from which they could be retrieved immediately, clients were assured, via encrypted transmission of an electronic scan or bonded courier delivery.

The ongoing (and interminable) process of digitizing historical analog records has created new records-management challenges for every institution—from major corporations to cultural institutions to individual families. The more we digitize, the more the backlog seems to grow. A portion of that backlog is in compromised condition, too fragile for a standard scan. And even as analog records acquire digital surrogates, the original documents often retain value and so stay on the shelf. Historian of photography Estelle Blaschke echoes many other archivists and historians in arguing that "the artifact does not become obsolete with … digitization"; "digitization is not preservation, but rather a safety backup."[19] That backup must then be continually backed up, and migrated as storage formats evolve. Thus, the digitized record must be remade over and over again. Data privacy and security present additional challenges. Recognizing the precarity of the digital record, many organizations have adopted redundant storage policies.[20] As they say, Lots of Copies (in distributed locations) Keep Stuff Safe!

Today, our information repositories face new external threats, too. Global warming and other anthropocenic natural disasters threaten to flood the basement storage rooms in our coastal cities' corporate headquarters. Fortunately, as records-management purveyors remind us, their subterranean facilities are "miles from danger zones associated with oceans, rivers, seismic fault lines, flood plains . . . [and] volcanic hot spots"; immune to airplane crashes, tornadoes, hurricanes, wildfires, and lightning; and, thanks to the presence of self-contained power and water sources and abundant on-site generators, insusceptible to brownouts and blackouts.[21] We're safe here from geologic and climatic threats. The greatest danger, we are told, is the archived media themselves, composed of highly flammable materials whose chemical degradation makes them even more incendiary. It's best, then, to retard that degradation in a carefully controlled subterranean climate. For extra precaution, facilities also separate storage units with

---

14   The Federal Reserve bunker also included a cold-storage area for those bodies escaping continuity of life. The Center for Land Use Interpretation, "The Nation's Media Archive: Taking Our Present into the Future," *Lay of the Land Newsletter*, Winter 2013, http://clui.org/newsletter/winter-2013/nations-media-archive; Michael Gaynor, "Inside the Library of Congress's Packard Campus for Audio-Visual Conservation," *The Washingtonian*, May 9, 2011, https://www.washingtonian.com/2011/05/09/inside-the-library-of-congresss-packard-campus-for-audio-visual-conservation/; Matt Novak, "The Fed's Cold War Bunker Had $4 Billion Cash for after the Apocalypse," *Paleofuture*, April 24, 2015, http://paleofuture.gizmodo.com/the-feds-cold-war-bunker-had-4-billion-cash-for-after-1699204253.

15   Library of Congress, "Library of Congress Packard Campus for Audio-Visual Conservation," http://www.loc.gov/avconservation/packard/.

16   Stephen Graham, *Vertical: The City from Satellites to Bunkers* (Brooklyn: Verso, 2016), 355. See also Luke Bennett, "The Bunker: Metaphor, Materiality, and Management," *Culture and Organization* 17, no. 2 (March 2011): 155–73.

17   Iron Mountain, http://www.ironmountain.com/. See also the Iron Mountain promotional video: "We Are Iron Mountain," http://www.ironmountain.com/Knowledge-Center/Reference-Library/View-by-Document-Type/Demonstrations-Videos/Tours/Company-Overview.aspx. Underground Vaults & Storage offers similar services and proudly notes on its website that it has received "AAA certification from the National Association for Information Destruction (NAID)." (Underground Vaults & Storage, "Document Resources Division: Professional Destruction Services," http://www.undergroundvaults.com/document-resources/.)

18   Bennett, "The Bunker," 169; Mike Gane, "Paul Virilio's Bunker Theorizing," *Theory, Culture & Society* 16, no. 5/6 (1999): 85–102.

19   Estelle Blaschke, "The Excess of the Archive," in *Documenting the World: Film, Photography, and the Scientific Record*, ed. Greg Mitman and Kelley Wilder (Chicago: University of Chicago Press, 2016), 242. See also Dan Nadel, "Burying the Past," *Metropolis Magazine*, November 2002, http://www.wilhelm-research.com/corbis/MetropolisMag_Nov_02_Corbis.pdf.

20   Cold War analog record managers also adopted "duplicate and disperse" strategies, "making copies of a document, often through microfilming, and shipping the copies to several distant locations with the hope that at least one would survive a nuclear attack" (Spencer, "Rise of the Shadow Libraries," 148).

thick firewalls, employ a number of automatic fire-suppression systems, and host on-site fire brigades.

But some disasters are inevitable. Iron Mountain facilities have suffered a number of fires over the years, including a fatal 2014 blaze in Argentina, allegedly set by tenants hoping to suppress evidence of tax fraud and money laundering. Records, despite their seeming banality, are a target for hostile activity. A resurgent Russia, with its skilled hacker corps, and a capricious, calamitous regime in the White House have revived many of our old Cold War suspicions about national and cyber security, and distributed acts of terror and civil unrest have threatened many global archives and cultural artifacts.[22] The Islamic State has famously destroyed numerous monuments, museums, and libraries in the Middle East. Fortunately, we're reminded, underground storage facilities are far from any likely military targets and urban sites of civil disruption. The mines tend to have a single entrance and exit, which facilitates monitoring by around-the-clock guards and surveillance video. They feature secure vault doors, redundant infrared and biometric security systems, and multiple alarms. And at Underground Vaults's salt mine in Hutchinson, Kansas, the only access is via a 650-foot drop, traversed by elevator; no getaway car can get away from there.

Iron Mountain even offers a potential defense against insurgent destruction elsewhere in the world. They've partnered with CyArk, a nonprofit dedicated to scanning and modeling global cultural heritage sites, to develop a comprehensive data management and archiving plan. The organization's data is saved on two magnetic tapes, one of which is delivered by Iron Mountain's secure MediaCare™ transport service, then stored in one of the company's underground tape vaults, where the materials are of course secured via carefully monitored temperature, humidity, and particulate filtration levels.[23] What does it mean to have a 3D model of an ancient ruin—say, the Mosul library or the Temple of Bel in Palmyra, both destroyed by ISIS—housed in a salt mine under Kansas? Some digital archaeologists propose that such archives of the ancient world might offer us the opportunity to recreate it, to undo its destruction.[24] Those heartland mines that once yielded physical building materials—stone and iron—can now yield architectural plans and digital scans, data-maps of the above-ground world, as it once was and perhaps could be again. The mine becomes a font of hope and regeneration.

Yet it is important to note that mines are still sites of violence and conflict, of tension between subterranean and above-ground political-geographies. Especially in the developing world, workers labor in harsh conditions, and locals are exposed to grave environmental hazards. Through a regime of "extractive imperialism," the exploited "colonial underground" feeds the insatiable appetites of the West.[25] Desires for land, labor, natural resources, and religious conversion drove enterprising nations into the Global South more than a half-millennium ago, and many of those same desires now draw global geoengineering corporations into the same regions. Still today, "the undergrounds of countries and continents are . . . being remade as volumes of postcolonial sovereignty, based on legal agreements that parcel them out to global mining firms under the armed protection of state security and paramilitary forces," writes Stephen Graham.[26] In the active mines that continue to fuel the global economy, there is little concern for national sovereignty, natural conservation, or the preservation of human life and heritage.

And in an interesting twist, many of our contemporary mining engineers—*data*-miners, the Silicon Valley elite—are retreating to the underworld. A growing "prepper" or survivalist community within the tech industry has invested in the construction of luxurious bunkers and vaults to escape impending environmental apocalypse or civil disorder. The latter unrest may be attributable, at least in part, to "a backlash against Silicon Valley," its job-killing robots, and its concentration of wealth.[27]

## Geological, Chemical, and Temporal Transformations

Particularly before regulation and reform, mines inflicted their "unflinching assault" upon labor and landscape in the West, too.[28] In the US, we still wrest coal, natural gas, stone, iron ore, limestone, sand, gravel, diamonds, gold, copper, silver, nickel, uranium, and other materials from our hills. But in some cases, those operations are sufficiently tamed and aestheticizied to allow for display—for on-site mining museums with gift shops and visitor tours. In Hutchinson, Kansas, extracted salt, incoming and outbound records, and visitors to the Strataca mining museum enter and exit through the same opening in the earth.

---

21   Underground Vaults & Storage, "Underground Storage: Disaster, Deterioration, and Deception Protection," http://www.undergroundvaults.com/offerings/secure-storage/underground-storage/. See also D. C. Hughes and V. J. Ryan, "Possibilities for Archives and Other Safe Storage Underground," International Society for Rock Mechanics International Symposium, Rockstore 80, Stockholm, June 23–17, 1980; and Tom Benjamin, "Adaptation of Underground Space," *National Archives* (March 1999), https://www.archives.gov/preservation/storage/underground-facilities.html.

22   Consider DataRefuge's efforts to preserve federal data—regarding climate and environmental research, in particular—that are believed to be particularly vulnerable during the Trump administration. See https://www.datarefuge.org/.

23   Iron Mountain, "Preserving the World's Heritage," 2013, http://www.ironmountain.com/Knowledge-Center/Reference-Library/View-by-Document-Type/Case-Studies/C/CyArk.aspx. For more on Iron Mountain's use of tape-based data storage, see Michele Hope, "Is Using Tape a 'Cool' Choice? The Numbers Tell the Story," Iron Mountain Knowledge Center, http://www.ironmountain.com/Knowledge-Center/Reference-Library/View-by-Document-Type/General-Articles/I/Is-Using-Tape-a-Cool-Choice-The-Numbers-Tell-the-Story.aspx.

24   Digital archaeology and the proposed recreation of destroyed cultural heritage sites are not without controversy. See Simon Jenkins, "After Palmyra, the Message to ISIS: What You Destroy, We Will Rebuild," *The Guardian*, March 29, 2016, https://www.theguardian.com/commentisfree/2016/mar/29/palmyra-message-isis-islamic-state-jihadis-orgy-destruction-heritage-restored; "Upon Reclaiming Palmyra, the Controversial Side of Digital Reconstruction," *NPR Weekend Edition Saturday*, April 2, 2016, http://www.npr.org/2016/04/02/472784720/upon-reclaiming-palmyra-the-controversial-side-of-digital-reconstruction.

25   James Petras and Henry Veltmeyer, *Extractive Imperialism in the Americas: Capitalism's New Frontier* (Leiden: Brill, 2014).

26   Graham, *Vertical*, 368.

27   Evan Osnos, "Doomsday Prep for the Super-Rich," *The New Yorker*, January 30, 2017, http://www.newyorker.com/magazine/2017/01/30/doomsday-prep-for-the-super-rich.

28   Mumford, *Technics and Civilization*, 69.

Where we once carted out rough blocks of limestone and chunks of salt, gouging ever-deeper into the earth, we today enact a new geological and architectural transformation: inserting steel doors and stacks, HVAC units and power cables, to turn caverns into vaults with geological properties conducive to the chemical preservation of material media. In these deep spaces of deep time, of rocks and minerals millions of years in the making, we strive to extend the lives of man-made artifacts that, above ground, in a more volatile climate, would disintegrate and decompose in a fraction of the time.

As Mumford explains, mines have long functioned as manufactured climates and time machines:

> Day has been abolished and the rhythm of nature broken: continuous day-and-night production first came into existence here. The miner must work by artificial light even though the sun be shining outside; still further down in the seams, he must work by artificial ventilation, too: a triumph of the "manufactured environment."[29]

In the repurposed mine, the mine-as-repository, bureaucracy meets climate and chemistry to reshape temporality and extend history.[30] These sites of extraction and conflict, of controlled geoengineering, are reborn as sites of controlled cultural preservation. In both cases, we exploit the landscape in order to regulate, and even reverse, organic material processes and the passage of time—to ensure our access to materials both produced and preserved across the *longue durée*.

Brian Michael Murphy poignantly notes that this "vast media preservation infrastructure," which "grew out of hauntings of destruction [and] fears of radioactive contamination," now reflects "our current fears, hopes, and persistent, impossible desires for permanent media invulnerable to the forces of (cyber)terrorism, natural disasters, and the indomitable force of decay that inheres in all media."[31] The slowness, stasis, and security of that subterranean repository are meant to provide a stable substrate for all the reckless speculation and rapid transactions transpiring in the world above. While the surface world accelerates, down below, we continue to dream of mastering geology and chemistry in order to engineer the retardation of time's passage. Our subterranean wager is that we'll find in the mines a secret to archival invincibility and perpetuity—as well as, perhaps, our own preservation.

Mined materials drive similar explorations in above-ground labs. Scientists and engineers are developing new storage media out of ceramic, nickel, tungsten, quartz, glass, even DNA, that promise to persist for thousands or billions of years.[32] Genomics and geology, in concert, might even provide a solution for the preservation and accessibility of our volatile digital records and obsolete digital media. In 2010 a group of European scientists, technologists, librarians, and archivists—the Planets group (Preservation and Long-term Access through NETworked Services)—created a time capsule containing five commonly used digital objects (a JPEG photo, an HTML website, a PDF brochure), along with file format specifications and encodings and a map of the "genetic code necessary to … access these file formats in the future."[33] That "digital genome," stored on an array of materials (flash drives, punch cards, microfilm, printed paper) to maximize accessibility to future generations, was then deposited in a metal box in the Swiss Fort Knox, a former bunker in the Alps that now houses high-security data centers. The experiment, according to Planets member Jacob Lant, of the British Library, was intended to "highlight the fragility of modern data, but also protect the tools for unlocking our digital heritage from a whole range of human, environment, and technological risks."[34]

It seems only fitting that mushrooms filled Knaust's mines before the files, films, and fiber optics arrived. Those natural decomposers foreshadowed our eternal struggle against time's passage and inevitable decay. While Iron Mountain's and SubTropolis's massive stone walls and sophisticated security systems might preclude the intrusion of moisture and gate-crashers, thus retarding the degradation of our data and documents, time still finds a way to sneak in. Over the *longue durée*, history itself takes its toll on our documents, the technologies we use to access them, and the individual minds and cultural ecologies in which those records hold significance. Yet we persevere in our preservation efforts, hoping, speculating, that these deep spaces of deep time will extend the lives of our human artifacts—and, by extension, human history and the human lives capable of remembering it—into a future of great uncertainty. Will anyone be around to mine the data?

---

29   Ibid., 69–70.

30   For more on the chemistry of archival storage, see the work of the Image Permanence Institute (https://www.imagepermanenceinstitute.org/index.php), and Paul N. Banks, "Overview of Alternative Space Options for Libraries and Archives," *National Archives* (March 1999): https://www.archives.gov/preservation/storage/overview-alt-space.html.

31   Brian Michael Murphy, "Bomb-proofing the Digital Image," *Media-N* 10, no. 1 (Spring 2014), http://median.newmediacaucus.org/art-infrastructures-hardware/bomb-proofing-the-digital-image-an-archaeology-of-media-preservation-infrastructure/.

32   "Eternal 5D Data Storage Could Record the History of Humankind," University of Southampton Press Release, February 18, 2016, http://www.southampton.ac.uk/news/2016/02/5d-data-storage-update.page#; Andy Extance, "How DNA Could Store All the World's Data," *Nature*, August 31, 2016, http://www.nature.com/news/how-dna-could-store-all-the-world-s-data-1.20496; Human Document Project 2014, http://hudoc2014.manucodiata.org/; Kevin Kelly, "Very Long-Term Backup," *The Rosetta Project*, August 20, 2008, http://rosettaproject.org/blog/02008/aug/20/very-long-term-backup/; Richard Kemeny, "All of Human Knowledge Buried in a Salt Mine," *The Atlantic*, January 9, 2017, https://www.theatlantic.com/technology/archive/2017/01/human-knowledge-salt-mine/512552/.

33   Lin Edwards, "'Digital Genome' Time Capsule Stored under the Swiss Alps," *Phys.org*, May 21, 2010, http://phys.org/news/2010-05-digital-genome-capsule-swiss-alps.html; Adam Farquhar, "Planets," *Digital Preservation*, http://www.digitalpreservation.gov/series/edge/planets.html; Planets, http://www.planets-project.eu/; Alasdair Wilkins, "Unlocking the Box That Holds the Secret to Digital Preservation," *Gizmodo*, June 11, 2010, http://io9.gizmodo.com/5558640/unlocking-the-box-that-holds-the-secret-to-digital-preservation.

34   Wilkins, "Unlocking the Box."

**Image Credits**
53–54: Courtesy of Connie Zhou

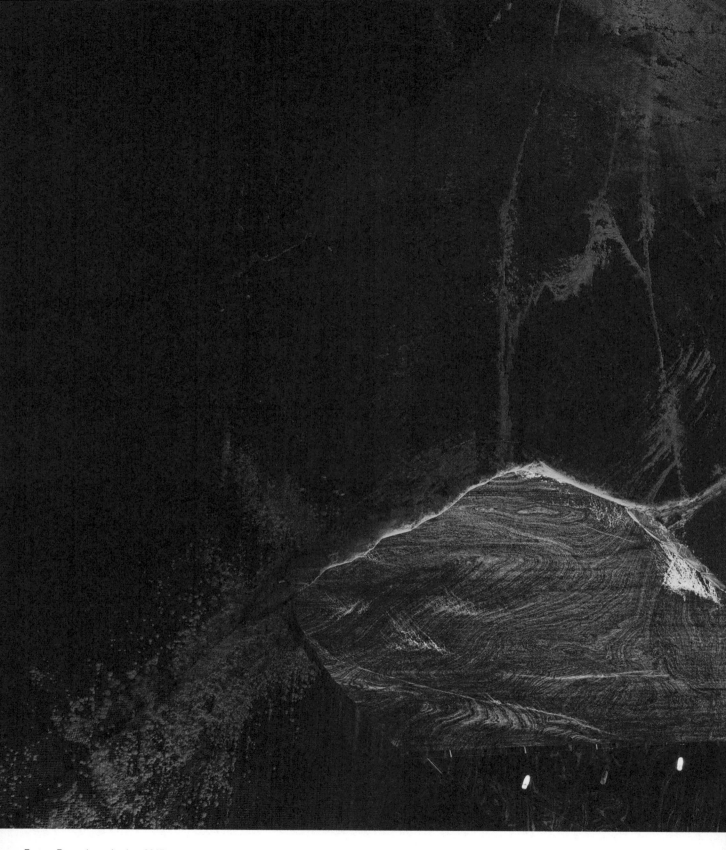

Former Rumanian salt mine, 2017.

Until now, we have not felt like underground dwellers because the natural system of the globe has seemed so large in comparison with any systems we might construct. That is changing. What is commonly called environmental consciousness could be described as subterranean consciousness—the awareness that we are in a very real sense not on the earth but inside it . . . the underground shelter is the emblem of humanity's ancient and honorable quest for truth, power, beauty, and security through technological achievements. As our burrows become more elaborate, however, the needs of the built environment may come to take precedence over the needs of the human builders . . . we should not forget that society too provides shelter, and in many cases a more flexible and effective kind. — Rosalind Williams, 2008

# Mapping the Future of Cities: Cartography, Urban Experience, and Subjectivity

**Antoine Picon & Carlo Ratti**

With the development of digital tools, urban cartography has undergone an array of spectacular transformations during the last few years. These transformations extend well beyond the spread of geographical information systems (GIS), both within city technical departments, and among designers and businesses responsible for the maintenance of buildings and infrastructure. Digital maps are everywhere. They pop up on computer and smartphone screens and punctuate urban life at various scales. The effect is a true revolution, one that corresponds to a more generalized technological shift and the advent of a new type of city, often described as the digital or smart city.[1] This turn is inseparable from the rise of a new urban subjectivity, the expression of which is far more dependent upon technology than ever before. Among its possible characterizations is the notion of the posthuman, envisaged as an emergent hybridization of humans and their technology. But such characterization is probably more misleading than enlightening in the early stages of this evolution—the ultimate consequences of which are far from clear. In addition, there are many ways to be posthuman besides appearing outwardly as a cyborg, the amalgamated organic and biomechatronic figure popularized by Hollywood action movies.

The effects of urban digital tools can be appraised at different levels. First and foremost, the cartographic revolution has meant the unprecedented diversification of the types of map that are available. Thanks to the increasing amounts of data produced by cities—often consisting of tracking data (that is, data that relates to specific points in physical space)—all sorts of phenomena can now be mapped, from environmental parameters such as pollution levels to the state of vehicular traffic. All around the world numerous urban research laboratories, such as the Massachusetts Institute of Technology (MIT)'s Senseable City Lab, are exploring new modes of visualization that will allow cities to be better understood and, above all, better managed.

Second, cartography—which has long been produced by trained specialists—is starting to be democratized. By using background maps supplied through platforms such as Google Maps, and by taking advantage of the ever-increasing quantities of data that are available over the internet, individuals and businesses can produce maps suited to their (or their clients') needs. OpenStreetMap, one of several shared mapping sites, is offering real competition to official geographical institutions. Mapping is no longer solely the preserve of experts. Its diffusion ranks among the factors that link a rapidly evolving urban cartography to the rise of a new subjectivity.

A third aspect of the changes that are afoot lies in the often dynamic or interactive character of on-screen maps. While traditional cartography produced representations that were fixed on paper, digital mapping allows the visualization of evolving situations. For example, we have all become accustomed to real-time maps of road traffic. Through these dynamic maps, the distinction between control and surveillance on the one hand, and cartography on the other, is blurring in parallel with the dividing line between reality and simulation.

The ubiquity of maps today reflects a profound transformation of digital culture that could perhaps be described as a "spatial turn," the term coined by geographer Edward Soja in the mid-1990s in relation to developments in the social sciences.[2] When it first emerged, digital culture tended to develop alongside physical space, without seeking to replace it. But under the influence of wireless network and geolocation technology, we have witnessed an increasing hybridization of physical space and digital content. The astounding rise of web-connected implements and consumer objects, which is generally referred to as the Internet of Things, is headed in the same direction: toward a digital culture that is profoundly spatialized. In this context, maps acquire a new dimension. It is not enough for them to plot out positions in space; they must also fix the point of convergence between atoms and bits. Such convergence will be significant to the transformation of cities in the coming decades.

## From Data to Maps

Have we not all, at some point, drifted into a daydream while studying a map or series of maps? Compared to other types of images, maps have the specific feature of being rooted in the world of data. A map is always the product of a set of information, the spatial distribution of which it makes legible. The relation between maps and data has taken an even sharper turn today with the exponential production of data across every last square inch of urbanized land. And because information is the ultimate asset in the emerging knowledge economy, the urban realm can be compared to an El Dorado that ever more companies are setting out to exploit.

A range of political and economic issues have surfaced around what is known as big data. First of all, to whom does information belong? Second, what scope should be allocated to specialist businesses that seek productive use of the enormous mass of data produced by cities? These problems are topics of keen debate. Even beyond questions regarding data ownership, there are others to be asked about who enjoys free access to it—the question of "open data." As a favored tool of data visualization, mapping is at the heart of these considerations.

Finally, how is the information displayed? With a change of color, or a different method of subdividing space to make the data relevant to it, a message can alter dramatically. Maps are never neutral—they are always the result of data manipulation, even if the computational power and well-designed interfaces of digital tools may seem to make them indisputable.[3]

---

[1] See, for example, Anthony Townsend, *Smart Cities: Big Data, Civic Hackers, and the Quest for a New Utopia* (New York: W. W. Norton and Company, 2013); Antoine Picon, *Smart Cities: A Spatialized Intelligence* (Chichester: John Wiley and Sons, 2015).

[2] Edward W. Soja, *Postmodern Geographies: The Reassertion of Space in Critical Social Theory* (London: Verso, 1989).

[3] See Laura Kurgan, *Close Up at a Distance: Mapping, Technology, and Politics* (New York: Zone Books, 2013).

## Dashboard, Commons, or Logbook?

As outlined above, several different types of urban mapping are in existence today. The first gives an account of the environment, and what is happening within infrastructures, streets, and buildings, in much the same way an observatory or dashboard would. In the years ahead, this type of map may also allow the actions of city residents, their state of health, and even their moods to be monitored more closely. This is likely to raise tough questions about respect for personal privacy.

This type of map is inextricably linked to the desire to manage cities better, as if the management process were somehow only a matter of steering. By the same token, the development of new tools for visualization (or simulation) of the urban realm is often linked to a powerful temptation to control. This could be described as neocybernetic, in the sense in which MIT mathematician Norbert Wiener first developed cybernetics in the 1950s and 1960s: a multidisciplinary approach to achieve better steering of complex systems.[4]

As a counterpoint to top-down oriented dashboard maps, a very different type of map has also gained in popularity. These maps intend to resist technocratic power and to promote more collaborative urban life. Some examples include activist maps, like the iSee mapping app that allows users to evade the surveillance video cameras installed in Manhattan, and others with similarly reflective purposes.[5] Within this varied group, it is worth highlighting the particular importance that maps assume as common property based on the collaborative accumulation of data or information exchange between people living in the same city. Today, maps may also serve the purpose of gathering suggestions regarding the future of the city. Implemented in a series of French cities, the *Carticipe* platform represents a new generation of participatory maps which are intended to foster citizen engagement and solicit public opinion. Of course, participation is dependent upon platform design and management techniques that may be more conducive to new forms of top-down government instead of enabling grassroots expression. But this risk is inherent to any form of collective mediation.

Beyond dashboards and commons, there are plenty of other uses for maps. For instance, maps can pinpoint individual or collective experiences, in the manner of a navigational logbook that mingles urban space with life experiences. Artists like Christian Nold or Sophia New and Daniel Belasco Rogers have explored this new type of use for maps, plotting the emotional content of certain urban locations, after the manner of the Situationists' psychogeographical investigations.[6]

From dashboards to commons to navigational logbooks, contemporary uses for cartography reflect diverse approaches to cities, favoring the construction of a shared information resource over a desire for rationalization or purely subjective experience of the urban realm.

## Toward the Smart City

To date, the physical form of cities—the arrangement of streets and squares—has only been marginally affected by the development of information and communications technology. Urban mapping, on the other hand, has undergone a genuine revolution. This reflects the real content of the transformations of the urban realm that are under way—the hybridization of atoms and bits, in the form of the smart city, represents a qualitative leap. This leap will manifest in our understanding of the city as a set of elementary occurrences, events, and scenarios, since it will soon be possible to monitor what is happening in real time, at every scale, and therefore to predict the trajectories most likely to ensue. For a very long time, cities were perceived as a collection of physical objects: a fortified enclosure with houses, civic buildings, basic infrastructure, and monuments. During the period of the First Industrial Revolution, cities were also assessed in terms of circulation and flow. Smart cities will continue to accommodate flows and networks, nested within a much wider scheme of events: billions of occurrences and overlapping situations that development scenarios will serve to regulate.[7]

Digital mapping appears as a preferred expression of the dynamics of the smart city's emergence, and as the medium through which some of its characteristics are firmly established. It reflects the real transformations of digital culture: a world populated by occurrences and events more than objects, and a mode of communication defined by the sending and receiving of messages more than any semantic configurations set in stone.

## Changing Urban Experience and Subjectivity

The rise of the smart city is inseparable from new forms of urban experience that merge geolocated data with the perception of a surrounding physical reality. Both tethered to a smartphone and immersed in city life, the new digitally augmented subject may be considered a cyborg, if one considers the vital character of its technological extensions, or prostheses.

But the figure of the cyborg presents a crude simplification of the way we relate to our environment through technological extensions. Following the intuition of cyberneticists like Gregory Bateson, we might instead conceive the contemporary digitally augmented subject as part of an ecology, as an entity almost inseparable from its physical environment.[8] Aren't we all, in a certain way, disseminated through various channels of the hybrid digital-physical world in which we live? This continuity is periodically challenged by the desire of the contemporary urban subject to regroup, or recoup force; to express his or her irreducible individuality. As an expression of this dual regime of dissemination and concentration, the

---

4   On the cybernetic approach, see Steve Joshua Heims, *Constructing a Social Science for Postwar America: The Cybernetics Group, 1946–1953* (Cambridge, MA: MIT Press, 1991).

5   Laura Kurgan and Eric Cadora, "Million Dollar Blocks," Spatial Information Design Lab, Columbia University Graduate School of Architecture, Planning, and Preservation (2006), http://spatialinformationdesignlab.org/projects/million-dollar-blocks.

6   "The Drawing of Our Lives," http://planbperformance.net/works/lifedrawing/; http://www.christiannold.com/. On Situationist maps, see Simon Sadler, *The Situationist City* (Cambridge, MA: MIT Press, 1998).

7   Picon, *Smart Cities*.

video-streaming feature Facebook Live approximates assimilation with the network, both revealing a singular point of view onto it—much like a digital age Leibnizian monad—while the simultaneous publishing of said view serves to assert one's allegedly unique identity.

We experience the contemporary city—the rapidly becoming smart city—in the same way. We may not yet be posthuman, but we do live partly beyond the boundary of our skins, with an extended perspective on the various events that punctuate urban life, and as unique islands of personal sensation, feeling, and judgment in the digitally augmented urban ocean, with its streams and swirls. Such a situation is not necessarily positive. There is a risk that this dual mode of existence may lend itself to new forms of domination by digitally diffuse corporate apparatuses that cater both to our extra-corporeal life and to our obsession with personal sensations. Is it any coincidence that a key advocate of the posthuman, author and futurist Raymond Kurzweil, is now employed by Google?

Maps do not provide an antidote to such risks. What is urgently needed at this stage is a critical cartography that can delineate more precisely the power of networked cities and their shortcomings, promises, and risks. Indeed, maps both reflect and serve the double life of the contemporary urban subject. Some chart our continuity with our urban surroundings, while others diagram our possible separation from it—think of the blue dot on a smartphone screen emphatically reminding that you are here, and not elsewhere, despite your capacity to roam freely. Maps foster innovative social links and personal records that have been made possible in the new digital culture regime. They enable us to share, but also to withdraw from the conversation in order to follow a personal itinerary. From commons to logbook, urban cartography draws the portrait of the contemporary urban subject, with all the ambiguities, tensions, and contradictions that implies.

**Singapore Land Transport Authority's (LTA) Intelligent Transport Systems Centre (ITSC).**

**Treepedia: a map of Singapore's urban tree cover.**

8   Gregory Bateson, *Steps to an Ecology of Mind: Collected Essays in Anthropology, Psychiatry, Evolution, and Epistemology* (San Francisco: Chandler Publishing Company, 1972).

**Image Credits**
Above: Courtesy of Land Transport Authority of Singapore

Below: Courtesy of Senseable City Lab, MIT

Tokyo Station, commuter rail, Japan, 2017.

It's easy, too easy, to depict the networked subject as being isolated, in contact with others only at the membrane that divides them. But if anything, the overriding quality of our era is porosity. Far from affording any kind of psychic sanctuary, the walls we mortar around ourselves turn out to be as penetrable a barrier as any other. Work invades our personal time, private leaks into public, the intimate is trivially shared, and the concerns of the wider world seep into what ought to be a space for recuperation and recovery. Above all, horror finds us wherever we are. — Adam Greenfield, 2017

# The Posthuman City:
# Urban Questions
# for the Near Future

# Alejandro Zaera-Polo

Since the 18th century, when the Western world became human-centered, neither humankind nor the very concept of the human has ceased to evolve. In 1933, Le Corbusier and a few other members of the International Congress of Modern Architecture (CIAM) issued *The Athens Charter*, a document aimed at orchestrating the emerging technologies of the built environment into a proposal for the future of cities.[1] A classification of human activities formed the vertebral spine of this proposal, structured around four urban functions: work, residence, leisure, and transport. This functional classification has structured urban planning policies ever since, but now its human-centered approach appears unable to address the problems of our age.

In the Anthropocene, humans have become capable of modifying natural ecosystems, geological structures, and even the climate; we have become so powerful that it is increasingly difficult to delimit the natural from the artificial. As the most populated human environment, cities are a central focus of these transformations, and yet none of these concerns seem to have permeated the tools used to plan cities. The urban planning disciplines are primarily conceived around human functions, despite the fact that the crucial issues they need to address—air pollution, rising water levels, drought, heat island effect, deforestation, biodiversity, food security, automated work, and inequality, to name a few—are primarily driven by concerns that, for the first time in history, transcend human societies and threaten the very survival of the planet. The economic, political, and technological drivers of modern urbanism—the mass integration of production, employment, and consumption; the separation of work, dwelling, recreation, and transportation; the division between the natural and the artificial—do not engage the urgent questions cities face today. Likewise, traditional urban instruments such as plazas, streets, and neighborhoods have now been privatized through neoliberal schemes and are ineffective in addressing the new urban collectives and constituencies, both human and nonhuman, which populate contemporary cities.

## Posthuman Cosmologies

Cities have enormous weight in the making of the Anthropocene. Witnesses of a veritable paradigm change, we must reformulate the cosmologies upon which the contemporary tools of urbanism have been constructed.

Arcane technologies and rituals of the urban were often based on mythological references. Ancient cosmologies were mechanisms of comprehension of the natural world that enabled different cultures to operate within the natural environment. The oldest ones predated human settlements and sought to explicate natural phenomena so as to regulate the modes of relation between humans and nature. As the urban environment became increasingly controlled by human agency cosmologies were discarded as systems of urban knowledge and governance. Typology and monumentality became primary tools for urbanism, with the structure of human relations prevailing over the physical and material determinations of the environment. The affairs of cities (*politika*) became an entirely artificial endeavor.

The current prevalence of the artificial and the political in the conceptualization of cities has tended to naturalize urban technologies while depoliticizing nature. However, pressing ecological concerns and the scale of technological development call for the imminent city to repoliticize both nature and technology, and to construct new urban cosmopolitics[2] which can support the development of new urban sensibilities.

An entirely new set of urban technologies has recently emerged, with the potential to radically transform urban protocols and experiences. Yet smartphones, GPS, electromobility, and biotechnology still remain largely beyond the expertise of urban planners and designers, who remain trapped in the humanistic precepts of modern urbanism.

Far from producing urbanity, urban functionalism has dismantled the commons and undermined urban democracy. Clichés like the relevance of squares or the percentage of public spaces as guarantors of community and urban democracy are as problematic as the inability of architects and urban planners to quantify the implications of density and urban form in either energy consumption or the determination of urban microclimates. The idea that architects and urban designers can locate effective agency in the distribution of human functions—such as work and residential—is at best naïve. Cities have become sources of extreme inequality and environmental degradation (in contempt of the *demos*, and all of the nonhuman urban constituencies too), which even threaten the subsistence of cities and point to the many unsurmountable contradictions at the core of current modes of economic integration. Theorists like Jeremy Rifkin and Paul Mason argue that we have already entered a post-capitalist world in which politics have shifted from the focus on capital and labor, to energy and resources.[3] Respectively, they have proposed new economies: shared economies of zero marginal cost driven by new technologies, and peer-to-peer organizations enhanced by pervasive computation, sustainable energy sources, and carbon-neutral technologies.

As the largest human habitat, cities have become the epicenters of global warming, air pollution, and a variety of ecological maladies. Naomi Klein has highlighted the fundamental opposition between capitalist growth and the limited natural resources of the earth, and questioned the capacity of capitalist regimes to resolve an imminent ecological catastrophe.[4] This has loaded ecology and technology—and more precisely, urban ecologies and technologies—with an

---

[1] Le Corbusier, Jean Giraudoux, and Jeanne de Villeneuve, *La Charte d'Athénes* (Paris: Plon, 1943).

[2] Albena Yaneva and Alejandro Zaera-Polo, eds., *What Is Cosmopolitical Design?: Design, Nature, and the Built Environment* (New York: Routledge, 2015).

[3] Jeremy Rifkin, *The Zero Marginal Cost Society: The Internet of Things, the Collaborative Commons, and the Eclipse of Capitalism* (London: Macmillan, 2014); Paul Mason, *Post Capitalism: A Guide to Our Future* (London: Allen Lane, 2015); and Paul Mason, "The End of Capitalism Has Begun," *The Guardian*, July 17, 2015, http://www.theguardian.com/books/2015/jul/17/postcapitalism-end-of-capitalism-begun.

[4] Naomi Klein, *This Changes Everything: Capitalism vs. the Climate* (New York: Simon & Schuster, 2014).

unprecedented political relevance. Cities today form a crucial entanglement between ecology, technology, and politics, where the equation of wealth, labor, resources, and energy must be reset to address the shortcomings of neoliberal policies.

## Imminent Commons

Does this scenario, determined by the Anthropocene casuistics and the crisis of neoliberal capitalism, imply that the work of urbanists and architects has been rendered futile? That the new commons will be entirely developed within social media? Has urbanism been expelled from politics entirely, so that it is now at the mercy of securitization and capital redistribution? On the contrary, some economists argue that urban planning, housing, and real estate hold the key to resolving urban inequality.[5] Cities precede the installation of political systems, and have systematically outlasted them, often constituting themselves in mechanisms of resistance to emerging power structures. For cities to become devices for the common good rather than instruments producing and implementing power structures—or inequality, or ecological destruction—urban practices must locate resources and technologies at their core. Rather than splitting urban life into functions easily captured by power, we should first identify the city's imminent urban commons and study how they might be reconstructed as instruments of devolution and ecological awareness, cutting transversally across technologies and resources. The following is an attempt to outline what those *imminent urban commons* might be, and how they may serve to revise urban practices.

## Ecology Commons

The first proposed revision to conventional urban cosmology resorts to the tradition of the primeval elements; by identifying the four commons related to natural resources, a new posthuman cosmology might be constructed. These four commons—air, water, fire (energy), earth—are driven by their various techniques of management. Each designated element has become crucially politicized, and therefore urbanized, by the ongoing environmental crisis, and is therefore of crucial concern to urban practices.

### Air

As Naomi Klein has stated, air is the element that most intimately binds all humans on earth together. While the ozone depletion from chlorofluorocarbons (CFCs) represented a significant and shocking threat, the effects of other airborne pollutants have been less conspicuous. Seven million people die every year from exposure to air-induced diseases.[6] Despite increased awareness of these problems, the air quality in cities seems to have gotten worse over the last few decades. According to a 2016 report published by the United Nations Children's Fund (UNICEF), "300 million children live in areas with extremely toxic levels of air pollution," and "approximately 2 billion children live in areas where pollution levels exceed the minimum air quality standards set by the World Health Organization (WHO)."[7]

Because cities are the densest sites of human population, attempts to clean air, channel polluted air away from city streets, prevent air stagnation, and improve airflow between buildings are central to preserving the rights to common natural resources of urban citizens. Since the early 20th century, buildings and cities have developed the ability to delimit, filter, and qualify air. However, as toxic emissions continue to rise each year, these abilities are becoming even more urgent and politically charged. While air is a universally needed resource, it is extremely difficult to quantify, visualize, sense, and model. Historically, this has been an obstacle to creating policies regulating its use. The emerging technologies of sensing and computational modeling will enable the development of more effective tools to manage the air commons.

### Water

Global warming and climate change do not only distort traditional climate patterns, but also reshape the earth itself. Rising sea levels are expected to have a massive impact on cities around the world in the coming decades, and their impact is already felt through natural disasters and plummeting real estate values. Urban waterfronts and riverfronts, traditionally key territories of urban life, are now threatened by climate change.[8] But the lack of access to clean water for urban populations is a growing humanitarian problem too. By 2050, one-third of people on earth may lack a clean, secure source of water.

Of the natural resource commons, none is so intimately tied to urban ecologies as water. Cities developed in the proximity of an adequate water supply, and the first urban policies appeared with respect to the use of water. In ancient Mesopotamia, the Code of Hammurabi included provisions for the distribution of water based on the area of fields farmed, dictated responsibilities for farmers to maintain canals, and assigned administrative responsibilities for canals. One primary responsibility of a city is to provide

---

[5] Matthew Rognlie, "Deciphering the Fall and Rise in the Net Capital Share: Accumulation or Scarcity?," *Brookings Papers on Economic Activity* (Spring 2015): 1–54; Thomas Piketty, *Capital in the 21st Century* (Cambridge, MA: Harvard University Press, 2014).

[6] World Health Organization (March 25, 2014) "7 Million Premature Deaths Annually Linked to Air Pollution" [News Release], http://www.who.int/mediacentre/news/releases/2014/air-pollution/en.

[7] Nicholas Rees, "Clear the Air for Children: The Impact of Air Pollution on Children," United Nations Children's Fund (UNICEF), October 2016, http://www.unicef.org/publications/files/UNICEF_Clear_the_Air_for_Children_30_Oct_2016.pdf.

[8] Laura Parker, "Climate Change Economics," *National Geographic*, February 2015, http://ngm.nationalgeographic.com/2015/02/climate-change-economics/parker-text.

infrastructure to deliver potable water to its citizens. Equally important to the health of a city is the ability to remove large amounts of water supplied by rains and flooding, but also produced by human waste. Cities function as the mediator between water-dependent organisms—that is, humans—and hydrological cycles driven by the evaporation, condensation, collection, and flow of water.

Centralized water purification and sewage treatment systems allowed modern cities to flourish, greatly increasing populations, human health, and the average life span. However, overreliance on centralized systems has begun to threaten water security. There is a threshold of efficiency for centralized systems: large amounts of energy are dedicated to sending water through extensive, corroded pipe networks. The decentralization of urban water infrastructure allows the city to become more resilient, for example, in the event of extreme weather conditions, power outages, terrorist attacks, hostile takeovers, government shutdowns, and other incidents.

Water retention, collection, and treatment systems are equally important areas of development for urban water infrastructure. The replacement of non-osmotic pavements (such as the asphalt surfaces that cover a large percentage of urban land), the introduction of bioswales (landscape design features that slow, collect, and filter surface water runoff), the recovery of buried streams within cities, and the development of water-retention systems for building envelopes are some current possibilities for urban practice that are likely to change the landscape of future cities.[9]

### Fire

Fire—the elemental placeholder for energy—is a vital issue that cities must address promptly. Fossil fuel consumption not only depletes natural resources, but more importantly, releases carbon into the atmosphere to cause climate change, which brings about pollution, health hazards, and rising water levels. Buildings consume 40 percent of global energy, 60 percent of global resources, and produce 48 percent of carbon emissions, with significant energy use concentrated in urban centers.[10]

The spatial distribution of energy sources, energy processing, and energy demand is a solid basis upon which different urban systems can be studied, historically or parametrically. Over time, urban energy systems have evolved progressively toward highly concentrated forms of energy, often obtained from fossil fuels, which can be easily transported. The development of fossil fuels as an energy source enabled deterritorialization, as portable forms of energy could sustain large and geographically dispersed industrial metropolises that would previously not have been possible without proximity to a naturally occurring resource.

The further development of solar, wind, tidal, and ground-sourced energy to power cities without resorting to fossil fuel combustion will profoundly alter the future urban cosmopolitics. As sustainable energy is primarily mediated through electricity, supplies will need to be driven locally, from various sustainable energy sources dependent on climate and geology. The ubiquitous fossil fuel resources will be phased out in the shift toward a reterritorialization of energies.

### Earth

The earth commons encompass the organic resources that make up the biosphere. The birth of urban civilizations was aligned with the ability to increase the productive capacity of land, in order to feed larger densities of people more reliably. Because most ancient citizens were farmers, there was a direct relationship between the productive capacity of the land, and the population that could be sustained in a city. This balance changed with advanced agricultural techniques and animal labor, and later, machines and synthetic fertilizers. Carbon footprints establish equivalences between land-measuring units and their capacity to produce energy from sustainable sources, to absorb carbon, and perform within hydrological cycles.

After a long exile from the city, plant matter, soil, and other organic elements are being reincorporated increasingly into city fabrics and building design. As a percentage of surface area coverage, organic matter affects such factors as reflectivity, humidity, thermal mass, heating and cooling cycles, and water runoff. Cities depend on the synthesis and resilience of life—human and otherwise. Urban land cannot be determined exclusively by functional assignments with respect to human behaviors: its productive capacity depends on topography, climate, and the bioactive layer of the soil.

Biotechnologies have developed increasingly effective urban applications. Some of the fastest growing projects include urban farming, hydroponics, and algae cultures with the ability to produce food, biofuels, and even illuminate cities. High-tech farming technologies such as hydroponics and aquaponics are constantly improved to increase efficiency and yield. The deployment of such technologies will become increasingly relevant to urban practices and assemblages.

---

9   Jeffrey D. Sachs, "The Challenge of Sustainable Water," *Scientific American*, December 1, 2006.

10   United Nations Environment Program, http://www.unep.fr/shared/publications/pdf/DTIx0916xPA-BuildingsClimate.pdf.

11   Marshall McLuhan, "The Gadget Lover," in *Understanding Media* (London: Routledge Classics, 2001): "With the arrival of electric technology, man extended, or set outside himself, a live model of the central nervous system itself" (page 53); and "It is to the railroad that the American city owes its abstract grid layout and the nonorganic separation of production, consumption, and residence. It is the motorcar that scrambled the abstract shape of the industrial town, mixing up its separated functions to a degree that has frustrated and baffled both planner and citizen . . . Metropolitan space is equally irrelevant for the telephone, the telegraph, the radio, and television. What the town planners call 'the human scale' in discussing ideal urban spaces is equally unrelated to these electric forms. Our electric-extensions of ourselves simply bypass space and time and create problems of human involvement and organization for which there is no precedent." (page 119).

# Technology Commons

The second series of commons that might constitute a new urban cosmology is related to the development of technologies that facilitate alternative forms of urban community. These commons present opportunities for new urban practices to pervade institutional, bureaucratic, and market mechanisms. Technologies can be considered extensions of human capacities, which have developed a posthuman life on their own. Or, as Marshall McLuhan writes in *Understanding Media* (1964), "... Technologies are self-amputations of our own organs."[11] The activities of sensing, communicating, moving, producing, and consuming have developed disembodied, collective forms which transcend their human origins.

### Sensing

The formation of urban sensibilities has a tradition that explores the city as an experience: the phenomenological focus of the late-19th-century City Beautiful movement and mid-20th-century Townscape theories approached the city as a sensual experience, as opposed to the more scientific, pragmatic approach of the Enlightenment. The emerging sensing technologies are opening entirely new possibilities: cities are experienced and sensed, but crucially, cities themselves sense. Weathervanes and watchtowers, common in traditional urban landscapes, provided citizens with weather forecasting and improved the navigation of increasingly complex urban spaces. Traffic wardens, firefighters, and other watchtower lookouts shared their advantaged sensibility with citizens in order to regulate urban processes. Weather patterns, security systems, and temporal rhythms have historically structured urban communities.

But these arcane sensing technologies have since become increasingly artificial, pervasive, and distributed. The proliferation of sensors in urban environments is one of the most defining features of contemporary urban milieus. Self-driving vehicles and other new urban technologies will not only provide new forms of mobility but also exponentially expand the population of sensing agents that document every physical feature, traffic delay, or change of air quality in the city environment. Computer vision in the form of light detection and ranging (LiDAR) laser scanners that collect point cloud data, high-resolution orthoimagery, and geographic information system (GIS)-enabled web applications enable us to visualize—and therefore act upon—urban processes that were previously inaccessible. When these sensors are interconnected, an unprecedented common, novel in its sensibility, will create a collective sensorium. When accessed via smartphone, open-source sensing data will be instantly accessible to urban populations, yielding constant updates regarding the urban environment.

Urban rhythms (such as alternating traffic lights, or subway train frequency) are increasingly controlled by algorithm, especially with respect to transportation, utilities, and security. In recent years, such dynamic regulation has widely expanded with the rollout of ubiquitous computing. Technologies such as city operating systems, closed-circuit television (CCTV) cameras, urban control rooms, smart grids, sensor networks, smart parking, smart lighting, city dashboards, and digital real-time information apps will not only have deep political implications in terms of surveillance and privacy, but also transformative spatiotemporal effects that will crucially alter the experience of cities. In effect, cities may have to be designed not only according to a human perspective, but through necessary interactions with nascent artificial sensibility.

### Communicating

One of the more powerful imminent commons is a consequence of the radical densification of urban communication networks, resulting from pervasive computation and wireless communications. While cities have always been characterized by dense communication networks—the postal service, the telegraph, the telephone—current technologies have intensified this drastically, through the combination of universal computing and the development of the World Wide Web, which allows users to interact and collaborate with each other as creative content-generators in a distributed virtual community.

In combination, the World Wide Web and wireless technologies have freed information entirely from physical attachments. The connection of mobile devices to global positioning systems (GPS) has allowed unprecedented opportunity to navigate urban space and engage with fellow citizens, creating entirely new urban geographies in the process. The possibility to connect with automated devices will certainly reshape the way in which we relate to our work or domestic environments. Platform capitalism deployed in urban areas—such as Uber, Lyft, Seamless, GrubHub—shows how some of these technologies are starting to affect urban life. And the development of new forms of domesticity, work, and leisure, and their impact on urban culture and politics, will disrupt traditional forms of urban functions.

Some of the most transformative processes triggered by new communication technologies relate to the possibility of sharing services and goods in time, including short-term residential rentals, bicycle- and car-share schemes, coworking spaces, and other initiatives based on shared economy. Inherited from modernist planning, today's city is primarily regulated by functional determinations and private property laws, which are being disrupted through shared ownership. Social media and universal computation hold the key to the development of new urban protocols, institutions, typologies, and experiences. Resource and energy management through sharing, recycling, or optimization is increasing, and current legislation must shift toward the normalization of sharing protocols. The urbanization of these technologies will open new potentials for architecture to engage with these emerging forms of urban culture.

### Moving

Human mobility has always been strongly influenced by technology, which has increased the radius of action far beyond the natural capacities of the human body. The origin of cities is inextricably linked to the development of collective forms of mobility and their infrastructures, which are able to supply the resources needed to sustain such concentrated populations that would otherwise be impossible without outreaching to a larger area where resources can be found to sustain the concentration of population required to form a city.

The radical increase in urban density caused by industrialization also multiplied the demand for mobility. The development of underground transit systems in London, Paris, Berlin, and New York was a crucial tool of urban development in the early 20th century, as these cities embraced industrialization as the fundamental mode of economic integration. In the post World War II period, the automobile transformed cities beyond recognition. Today, transportation is one of the most energy-intensive activities in the city, and currently accounts for a large percentage of overall carbon emissions and pollution.

In response to these ecological concerns, unipersonal electromobility and self-driving vehicles are likely to alter urban traffic patterns even further. Electric vehicles (EVs) and sharing schemes are now spreading worldwide, anticipating decentralized urban transport systems with a minimal carbon footprint—unlike previous modes of individualized transport—and breaking with models of vehicle ownership. Automated logistics are also an important part of urban mobility developments.

### Making

If the late capitalist city is characterized by the exile of production from the urban core, and the takeover of financial services as well as securitization through residential markets are now key components of urban economies, there are also new technologies for digital fabrication, laser 3D scanning, 3D printing, and robotics that have since relocated some high-value fabrication activities back to the urban core. The potential impact of the return of production—digitally enhanced, automated, etc.—to the city indexes the emergence of new concerns, such as the distribution of work and benefits and the relocalization of the production infrastructures within the urban fabric, which will require new governance protocols for the emerging urban *Homo faber*.

While contemporary urban culture has often been associated with the rise of the *knowledge society*, where most of the wealth and employment is created by the production of knowledge, contemporary economies have now reached a point where the progressive elimination of agricultural and industrial sectors is making urban economies untenable.[12] After decades of service-driven urban economies and real estate speculation, emerging fabrication technologies may be able to effectively reindustrialize cities.

The so-called Fourth Industrial Revolution (4IR) is the fourth major industrial era since the Industrial Revolution of the 18th century and is enacted by a range of new technologies that fuse the physical, digital, and biological worlds, with an impact on all disciplines, economies, and industries. Central to this revolution are emerging technological breakthroughs in fields such as artificial intelligence, robotics, the Internet of Things (IoT), autonomous vehicles, 3D printing, biocomputation, and nanotechnology.

The 4IR has also created its own urban cultures: makers practice a technology-based extension of DIY activities that intersect with hacking actions. The maker culture remains committed to the physical world, operating according to specific moral principles and producing distinctive forms of inhabiting and occupying urban space. Shared spaces for production and fabrication equipment and for processing recycled or reclaimed materials are key aspects for these emerging ethics of urban production, often implemented through small-scale urban operations. A revision of new urban production technologies and how they may be reinserted into the urban fabric is now an important field of consideration for the making common.

### Recycling

Metropolitan governments have paid increasing attention to the collection, sorting, and recycling of urban waste and biosolids. These processes have reached a geopolitical dimension, with regional and even transcontinental systems set out for treating refuse. The cultural dimension of recycling protocols is substantial and directly impacts citizens and their perceptions of the city. For decades, Japan has been at the forefront of fostering collective consciousness of recycling, and installing infrastructure to optimize urban metabolism. Many cities have attempted to follow the Japanese example, only to realize the cultural difficulty of implementing such programs. Installing multiple-track recycling protocols in contemporary cities appears to be an insurmountable cultural problem, and yet robotic sorting of solid and electronic waste is a real alternative to landfill and must be expanded to urban areas lacking a Japanese cultural background.

Human waste is also a subject of increasing attention for municipalities. A more sophisticated approach to its recycling could have an enormous ecological impact by reducing global dependence on synthetic fertilizers. In the light of recent interest to promote urban farming for the purposes of food security and $CO_2$ absorption, for example, the recycling of human waste obviously has great potential. Waste composting for urban farms on roof gardens, private courtyards, and bioswales appears as the natural destination of human waste on a local, granular urban scale. As Pierre Bélanger has stated, "Waste is the 21st-century food."[13]

## Ecologies Rather Than Functions

It appears inevitable for urban practices of the immediate future to incorporate the emerging technologies discussed above in fields where the new urban commons are to be found, ranging from governance to production. Especially because urban planning based on human functions has now become a mechanism to divide and wield power rather than produce urbanity, new instruments must be developed to address the imminent posthuman commons.

---

12   Ha-Joon Chang, "Making Things Matters: This Is What Britain Forgot," *The Guardian*, May 18, 2016, https://www.theguardian.com/commentisfree/2016/may/18/making-things-matter-britain-forgot-manufacturing-brexit.

13   Pierre Bélanger, "Landscapes of Disassembly," *Topos* 60 (2007): 83–91.

We live in an age marked by vast technological developments that must be better incorporated into the ways we conceive and design cities within wider global ecologies. The ancient elements of air, water, fire, and earth along with artificially enhanced sensibilities, collectivities, logistics, and metabolic processes enabled by emerging technologies should become the central concern of a posthuman urbanism, focused on the ecologies and economies of elements and milieus. In posthuman urbanism, cities are designed to engage with concerns much broader than the organization of delivery systems responsive to human activities, as if these infrastructures or needs were independent of the milieus, tools, climates, and topographies in which they are situated. An upgrade from the modernist, human-centered functionalities—and the expired sensibilities associated with them—that undergird contemporary urban practice is now urgently needed.

Coal mine, Mongolia, 2015.

[F]rom gold to gravel, copper to coltan, iron to uranium, geological resources represent the invisible mineral media—below the visible surface of the earth—that supports technological aspects of so-called modern life. In subway tunnels or on suburban streets, in electronic manufacturing or information media, on stock exchanges or in commodity markets, the geological materiality of contemporary urbanism may seem inescapable, but the marks of technological imperialism seem even more indelible. . . . Where do these materials come from? Who do they belong to? Under whose jurisdiction? How are they moved and removed? Where do they go? Who processes them? What energies are required? What do they leave behind? — Pierre Bélanger, 2016

# Beyond the Threshold of the Human

# A Conversation with Eyal Weizman

# Mengele's Skull and the Paradigm Shift in Forensics

| | | |
|---|---|---|
| | New Geographies | We came across a small book of yours, *Mengele's Skull: The Advent of a Forensic Aesthetics* (2012), which might be a good entry point to our conversation.[1] |
| Eyal Weizman | | Thomas Keenan and I published it together. *Mengele's Skull* marks an important moment in the conceptualization—and finally, the birth of—Forensic Architecture. It describes a significant change in conflict analysis, a shift from relying on testimony to the materiality of evidence (represented in this case by the skull). It is a shift toward a particular kind of osteobiographical evidence. |
| | NG | Could you briefly describe what the book is about? |
| EW | | Josef Mengele was a Nazi perpetrator and a doctor at Auschwitz who escaped Germany after World War II. He lived out his life hiding in Argentina until his death in 1979. His body was later exhumed in Brazil in 1985. The ensuing process of identification was a turning point—in legal and technological terms—that inaugurated a new forensic sensibility. Mengele's skull was discovered in Embu das Artes, a small suburb of São Paulo. It was analyzed by Brazilian, German, and American forensic anthropologists and other experts. They developed a new technique that changed the focus from evidence provided by oral testimony and witnessing—in cases of genocide and other abuses of human rights—to a *nonhuman* approach based on osteology and osteobiography (the biography of bones). This led to the emergence of an important organization in Buenos Aires, the Equipo Argentino de Antropología Forense (EAAF), or Argentine Forensic Anthropology Team.[2] Their work throughout the 1980s enabled a paradigm shift in human rights advocacy and research, which expanded from an almost exclusively testimony-based practice to include a host of scientific practices attuned to the texture and form of bones and minerals, among other things. By developing and implementing a new scientific method, the EAAF was able to recover the bodies of many people who were disappeared or killed during Argentina's military dictatorship.[3] For those of us working in the field of Forensic Architecture, the advances introduced by the EAAF during this time established a critical standard in the context of war crime investigation that radically broadened the spectrum of evidence to incorporate material inference. We are lucky to have been able to collaborate closely with the EAAF, whose scientific work has done amazing things for human rights as well as for international law. |
| | NG | How is this work related to your approach as an architect? |
| EW | | My personal interest in the form and texture of bones is connected to my sensibility as an architect, which is very much concerned with investigating form and matter. For me, architecture in effect is *matter in movement*, whose traces in territorial textures and forms allow the reading of political processes. |
| | NG | When did you become involved with the EAAF? |
| EW | | As a student, I read all about the work of the Argentinian forensic anthropologists, but *Mengele's Skull* is a relatively late tribute to the group. In many respects, the book was essential to the start of our Forensic Architecture project.[4] About a year ago we began working together with the EAAF on a major investigation of the disappeared bodies in Mexico, which connected their skill in exhumation and identification with ours in urban analysis. |
| | NG | How interesting that the EAAF played such a fundamental role in the development of Forensic Architecture. We see some important insights about it in your book *Hollow Land: Israel's Architecture of Occupation* (2007) too, albeit in embryonic form.[5] |

EW    *Hollow Land* was developed from my PhD, which was titled "The Politics of Verticality." It is also a very personal project. The entire book was written out in notebooks during the Intifada,[6] as I was drawing a map of the West Bank. The situation was rather precarious then, and we needed to be weary of shots coming from both sides when passing through that area.

## Human Rights beyond the Human

NG    The question of rights is a very important one, particularly from a posthumanist point of view, because it remains essentially bound to a particular conception of the "human"—that of the "liberal subject."[7]

EW    When thinking about posthumanity and the posthuman, in principle, one assumes that human rights is the last field to benefit from those insights, because its ethical metrics are so attached to the figure and the ethics of the "human." Yet despite this initial impression, on further examination we see that the field of human rights is continuously transforming—ethically, epistemologically, and politically. If you think of the post-1970s phase, as opposed to the postcolonial phase of the 1950s, for example, human rights gave voice to individuals facing political repression—as in the case of dissidents in the Soviet Bloc rising up against the totalitarian state. Within this historical context, "human testimony" and "memory" serve not only as epistemological resources but as ethical substance, too. They provide critical evidence—what people remember when describing their own traumatic experiences—and pitch a human voice against the arbitrary, brute power of the state. In this sense, human rights, both as a political ethic and as a methodology (with its associated techniques of interviewing, recording, and transcribing victim testimonies), can be said to be "humane." But starting with the work of the EAAF in Argentina in the 1980s there was a split—after which the methodology became scientific, and by so doing, went beyond the limits of the living human voice. The ethico-political aim, of course, remained anti-totalitarian—in that particular case against the military junta, and the oppression and terror people suffered under it. During the *Juicio a las Juntas* (Trial of the Juntas), the deputy prosecutor Luis Moreno Ocampo[8] presented bones for the first time in court as evidence, thus enabling human rights to decouple, methodologically, from the humanism of its purpose. The aim remained humanist, but the method turned beyond it.

NG    In what sense do you think the methodology turned beyond a humanist framework?

EW    One of the assumptions of human rights—in various religious theologies, but Christianity in particular—is that the person who has witnessed an atrocity or experienced torture can describe that pain in language: as testimony. The victim is considered a certain kind of messenger, almost in a religious way. But testimony is more than an epistemological resource. Its meaning lies beyond what a person says—it implies that speech, as testimony, has value simply by virtue of being given, that the human voice has political significance in itself. What happened in Argentina is that anthropology, chemistry, odontology, and various other sciences suddenly entered, as voices, into the field of human rights. That, in turn, introduced a series of fundamental questions: Can those landscapes in which atrocities and other crimes have taken place—along with their architecture, their infrastructure, the plant life that grows there, the insects that populate their fields—be understood as historical records, specifically, as records of *violence*? In the texture and form of the human remains retrieved from such sites, how can we read the markings of historical events? It is almost like a shift from history to archaeology, in the sense that you confront materiality without language, a form of evidence without speech or a written record to describe it. As a result, the field of forensics created an opening—it allowed various types of sensibilities to enter into contact with traditional human rights discourses.

To get back to your question, finally, we might ask: As a discursive practice or a political ethics, can human rights remain the same after its forensic turn? Is science only a tool used to understand, record, and document the events of the past? Or does the fact that we can now look at landscapes, buildings, bones, images, data, and code as material evidence open up the field into things that are *beyond* the human? In other words, if the methodology and the epistemology of human rights discourse changes, do its political ethics? I would argue that yes, indeed, this change enabled various other shifts to take place.

NG   To dwell for a moment on the emergence of forensics as a material-based practice comparable to archaeology, what is so compelling about the recent *Forensic Architecture* case study of the Naqab/Negev desert is the attempt to rewrite the territorial and anthropological history of that region by means of "counter-forensic" types of evidence, from indexing droughts and travel memoirs, to the legal basis for land tenure. How do you approach different scales of historical time in your research?

EW   That particular study came out of a very basic question: How do you write the history of an area—here, the Naqab desert—not only in terms of the displacement or destruction of people and their villages, but also in terms of environmental and climatic transformation? In all the books written about the history of the Naqab, this element is not really discussed. How do you write climate and the environment into a human rights framework? We often have a fixed idea of what human rights are, and we tend to associate them with classic formats such as the Truth Commission.[9] But when you put the problem in terms of writing a history of the climate, how can you articulate a new sort of "truth commission" that would take into consideration such historical unfolding? I would argue that you need to look at a range of "intersections of evidence": the crossing of testimonial, material, osteological, mineral, and even literary evidence. . . . In this sense, I needed to develop a story that was simultaneously deep and wide. Deep—that is, going back in time not only to the Byzantine occupation of this area, but further still to the fluctuating aridity line of the prehistoric period. And extremely wide—that is, comparative along an extensive string of territorial connections, from North Africa all the way to Iran, Pakistan, Afghanistan, and so on. That is precisely the idea behind forensics; that you can trace, from one bit of material reality—one village, for example—how historical periods have "mineralized" into stone, and then find connections running along a wide territorial equator and moving across this conflict-ridden land. All Forensic Architecture cases can be considered singular examples (e.g., one village that has been evicted 116 times). We are interested in addressing precisely how to read in these monadic cases a history of the *longue durée*. In one of my current projects, a new book, I'm looking into the notion of the "split second." In other words, those split-second decisions which can be said to contain the entire history of slavery and colonialism, and in which different times coexist within certain singular moments.

# Elastic Space

NG   What you describe has relevance to your research methods. On the one hand, we see your work as an attempt to describe a new kind of territorial complexity, one in which architecture is not defined through a Western epistemological lens (i.e., buildings are passive objects to be designed), but as a set of "spatial and political technologies" that play out in a larger three-dimensional field; an "elastic frontier" articulated by the interactions and entanglements of spatial forms and human as well as nonhuman bodies, unfolding against a background of specific political, social, and cultural dynamics. On the other hand, so as to substantiate this approach, you have come up with a very precise language to conceptualize, map, and otherwise express these forms of spatiality.

EW  Maybe you are right. To pick up your point on the "elastic frontier" and enter the topic from there, I'll note that when thinking about the elasticity of space, you cannot avoid the force-form relation. Form is always an embodiment of forces, and it is indeed possible to read forces in terms of form. The idea of elasticity allows for the continuous rearticulation of forces, which converge into specific morphologies. The challenge of this method is that the reading of such forms must keep pace with their transformations. To give an example: the problem we confronted when analyzing the forensics of the West Bank was that we needed to describe a process that was still ongoing, and transforming very rapidly. While I was there, I could see the environment changing continuously in front of my eyes, wherever I went or looked. You can't stop or fix that process in order to conceptualize it. On the contrary, your method needs to mirror a movement that is always intrinsically political—politics are, effectively, *matter in movement*. My contention is that the reorganization of the landscape is so fundamental, and at the same time so diagrammatic, that one can actually read politics from form, and do so at several speeds and at various scales. For instance, if you look at the history of the West Bank and perform the mental exercise of playing out five years of territorial and environmental change in five minutes, you will be able to see the way in which politics is fossilized into different formal and spatial organizations. If you now freeze that mental animation, you can read how politics was transformed into matter, frame by frame. In other words, how political forces shifted this or that pile of matter across the terrain. This idea is close to Rob Nixon's notion of "slow violence,"[10] as different events take place on the scale of days, weeks, months, and years. This same process is manifested when there is an impact and something blows up, or a piece of a building is destroyed—that is, the form-force relationship can also be seen in those instances. Elasticity implies that there is no fundamental difference between construction and destruction. We are always talking about the transformation of matter across the surface of the earth, about a diagram of continuously changing forces, arranged in a multiplicity of forms.

## Thresholds

NG  The concept that fascinates us most, and that is omnipresent in your work, is the "threshold." Indeed, many of the cases presented in *Forensic Architecture* (2017) unfold at the intersection of multiple thresholds, which you have also referred to as "spaces of political void," "frontier regions," or "border regions." It appears that in these thresholds the law, norms, and the state or international code of conduct become plastic, so to speak. The threshold gives space to various forms of offense; it is a gray area between the legal and extralegal. In this context, it is striking to see a certain paradox at play: while the design disciplines involved in the formation of the built environment (architecture, urban planning, landscape) are typically mobilized under a humanitarian rubric, they are often instrumental in reinforcing multiple forms of violence and oppression, by both state and extrastate actors.

EW  I would like to think of *Forensic Architecture* as something like a detective novel, a format which always contains two entangled plots: one in the past (the crime) and one in the present (the investigation). I hope my next book on the split second, which I mentioned earlier, will read more like a thriller. It's about the shrinking down of historical time: from the historical account of the long conflict between Palestine and Israel, as outlined in *Hollow Land*, to what happens within the split-second span of incidents. I am trying to show how these two durations are related and how they interact; that is, how you can conceptually visualize the consequences of the *longue durée* in a split second, and vice versa.

Now, to address the question of thresholds—what interests me is how different thresholds relate to each other: the threshold of the territory, the threshold of detectability, the threshold of the law. Such specific limits come together in manifold ways. In Waziristan,[11] a case analyzed in *Forensic Architecture*, all these forms of threshold converge. It is an area that is outside the law (it has been a zone of exception since the time of the British Raj), beyond visibility, and behind siege lines. These conditions are what enables, among other things, drone warfare to take place there.

In my recent essay, "Are They Human?" (2016),[12] on the cultural history of the orangutan, yet another threshold comes into view—the threshold of the human. There I discuss an orangutan's trial case in Argentina. In 2014, an Argentinian court ruled that Sandra (the orangutan) was a sentient being with thoughts and feelings and that she was a "nonhuman person" who had been wrongfully deprived of her freedom in the zoo. The case of Sandra poses many important problems: where does the human end and where does the nonhuman begin? The orangutan has been sitting on this very boundary line for the last 250 years. But on which side is she—in or out? Both our conception of humanity and the notion of human rights are deeply related to a fundamental question: How can we account for what being "human" is?

But this topic is very complex and perhaps deserves another conversation.

1   Thomas Keenan and Eyal Weizman, *Mengele's Skull: The Advent of Forensic Aesthetics* (Berlin: Sternberg Press, 2012).

2   The Argentine Forensic Anthropology Team (*Equipo Argentino de Antropología Forense*, EAAF) is a nonprofit nongovernmental organization that applies forensic science (anthropology and archaeology) to the investigation of human rights violations in Argentina and worldwide. EAAF formed in 1984 to investigate the cases of at least 9,000 people disappeared by the military government that ruled from 1976 to 1983. See www.eaaf.org.

3   During the military dictatorship in Argentina (1976–1983), more than 30,000 people were sequestered, tortured, and disappeared by the government, which operated in collusion with local civil oligarchies.

4   Forensic Architecture "refers to the production of architectural evidence and to its presentation in juridical and political forums ... Forensic Architecture is also the name of a research agency established in 2010, together with a group of fellow architects, artists, filmmakers, journalists, and lawyers. ... The agency produces evidence files that include building surveys, models, animations, video analyses, and interactive cartographies, and presents them in forums such as international courts, truth commissions, citizen tribunals, human rights and environmental reports, and, on one occasion, in the UN General Assembly." Eyal Weizman, *Forensic Architecture: Violence at the Threshold of Detectability* (New York: Zone Books, 2017), 9.

5   "Architectural practice in conflict zones could ... incorporate the ethical motivations, and the methodological capacities, for bearing professional witness to those crimes conducted through the transformation of the built environment." Eyal Weizman, *Hollow Land: Israel's Architecture of Occupation* (London: Verso Books, 2007), 260.

6   The second Intifada (September 2000–February 2005) was the second Palestinian uprising against Israel that became a bloodbath over four years. During this time it is estimated that 3,000 Palestinians, 1,000 Israelis, and 64 foreigners were killed.

7   See especially Cary Wolfe, "Flesh and Finitude: Bioethics and the Philosophy of the Living," *What Is Posthumanism?* (Minneapolis: University of Minnesota Press, 2010), 49–98.

8   Argentine lawyer Luis Gabriel Moreno Ocampo came to prominence as assistant prosecutor in the Trial of the Juntas, which resulted in convictions for five of nine senior military commanders charged with mass killing, murder, kidnapping, and torture.

9   "A truth commission is (1) focused on the past, rather than ongoing events; (2) investigates a pattern of events that took place over a period of time; (3) engages directly and broadly with the affected population, gathering information on their experiences; (4) is a temporary body, with the aim of concluding with a final report; and (5) is officially authorized or empowered by the state under review." See Priscilla Hayner, *Unspeakable Truths: Transitional Justice and the Challenge of Truth Commissions* (London: Routledge, 2011), 11–12.

10  Rob Nixon, *Slow Violence and the Environmentalism of the Poor* (Cambridge, MA: Harvard University Press, 2011).

11  See Weizman, "Introduction: At the Threshold of Detectability," *Forensic Architecture*, 13–47.

12  Eyal Weizman, "Are They Human?" *E-Flux Architecture: Superhumanity* 51 (October 2016), www.e-flux.com/architecture/superhumanity/68645/are-they-human.

# Satellite:
# Enigmatic Presence

# Stephen Graham

We live in a satellite enabled age. The satellites flying above us are not abstract agents of science but part of the critical life support system we all depend on, every day.[1]

As one ascends into the sky above the earth's surface—departing from the largely upright human experience of living on it—there comes a point when the conventions surrounding the human experience of the vertical dimension must inevitably break down.

At the margins of the earth's atmosphere and the threshold of the vast realms of space we enter a world of orbits. Here we start to encounter the crucial but neglected manufactured environment of satellites and space junk.

"Verticality pushed to its extreme becomes orbital," multimedia artist Dario Solman reflects. At such a point, "the difference between vertical and horizontal ceases to exist." Such a development brings with it profound and unsettling philosophical challenges for a species evolved to live upright on *terra firma*. "Every time verticality and horizontality blend together and discourses lose internal gravity," Solman argues, "there is a need for the arts."[2]

The earth's rapidly expanding array of around 950 active satellites—over 400 of which are owned by the United States—are central to the organization and experience, as well as destruction, of contemporary life on the earth's surface. And yet it remains difficult to visualize and understand the enigmatic presence of this satellite array. Mysterious and cordoned-off ground stations dot the earth's terrain, with futuristic radomes and relay facilities directed upwards to unknowable receivers above. Small antennae lift upwards from myriad apartment blocks to quietly collect invisible broadcasts from transnational television stations. Crowds might even occasionally witness the spectacle of a satellite launch atop a rocket.

Once aloft, satellites become distant and, quite literally, "unearthly."[3] At best, careful observers of the night sky might catch the steady march of mysterious dots across the heavens as they momentarily reflect the sun's light.

Such a small range of direct experience with satellites fails to equip us with the skills necessary to disentangle the politics embedded in this huge aerial assemblage of circling and (geo)stationary bodies.

It doesn't help that the literature on satellites in the social sciences is startlingly small. Communications scholars Lisa Parks and James Schwoch suggest that this is because scholars, too, struggle to engage with satellites because they lie so firmly beyond the visceral worlds of everyday experience and visibility. "Since they are seemingly so out of reach (both physically and financially)," they emphasize, "we scarcely imagine them as part of everyday life"at all.[4]

The continued tendency of many scholars of geography and geopolitics to maintain a resolutely horizontal view compounds our difficulties in taking seriously the crucial roles of orbital geographies in shaping life (and death) on the ground. Only very recently have critical geographers started to look upwards to the devices circling earth in the first tentative steps toward charting a political geography of inner and outer space.

Such a project emphasizes how the regimes of power organized through satellites and other space systems are interwoven with the production of violence, inequality, and injustice on the terrestrial surface.[5] But it also attends to the importance of how space is imagined and represented as a national frontier; a birthright of states; a sphere of heroic exploration; a fictional realm; or as a vulnerable overhead domain through which malign others might stealthily threaten the societies below at any moment.

The invisibility of the earth's satellites and their apparent removal from the world of earthly politics has made it very easy to keep their organization and governance from democratic or public scrutiny. Widening domains of terrestrial life are now mediated by farabove satellite arrays in such fundamental ways that they have quickly been taken for granted—when they are even noticed or considered at all.

Global communication, navigation, science, trade and cartography have, in particular, been totally revolutionized by satellites in the last few decades. Military GPS systems, used to drop lethal ordnance on any point on earth, have been opened up to civilian uses. They now organize the global measurement of time as well as the navigation of children to school, yachts to harbor, cars into supermarkets, farmers around fields, runners and bicyclists along trails and roads, and hikers up to mountain tops.

Broader access to powerful imaging satellites, similarly, has allowed high-resolution images to transform urban planning, agriculture, forestry, environmental management, and efforts by NGOs to track human rights abuses.[6]

Digital photography from many of the prosthetic eyes circling above the earth, meanwhile, offers resolutions that Cold War military strategists could only dream of—delivered via the satellite and optic-fiber channels of the internet to anyone with a laptop or smartphone. A cornucopia of distant TV stations are now accessible through the most basic aerial or broadband TV or internet connection. Virtually all efforts at social and political mobilization rely on GPS and satellite mapping and imaging to organize and get their messages across.

Satellites, in other words, now constitute a key part of the public realm of our planet. The way they girdle our globe matters fundamentally and profoundly. And yet satellites are regulated and managed by a scattered array of esoteric governance agencies. They are developed and engineered by an equally hidden range of state and corporate research and development centers. When the obsessive secrecy of national security states is added to this mix, it becomes extremely hard

---

1   *Satellites for Everyone: The Big Picture, Satellites Application Catapult*, 2014, http://www.spaceforsmartergovernment.uk/workspace/assets/files/satellites-for-everyone-s4e-br-5404e13017bd5.pdf.

2   Dario Solman, "Air Attack," *Airfiles Blog*, 2001, http://filmlog.org/airfiles/dat-nav/t-s-htm/attf/attf07.html.

3   Jim Oberg, cited in Trevor Paglen, *The Last Pictures* (New York: Creative Time Books, 2012): 2.

4   Lisa Parks, "Orbital Performers and Satellite Translators: Media Art in the Age of Ionospheric Exchange," *Quarterly Review of Film and Video* 24, no. 3 (2007): 207–08.

5   An important book in this debate is Daniel Sage's *How Outer Space Made America* (Farnham: Ashgate Publishing Limited, 2014).

6   On the latter, see "Geospatial Technologies Project," Scientific Responsibility, Human Rights and Law Program, American Association for the Advancement of Science (AAAS), https://www.aaas.org/page/geospatial-technologies-project.

to pin down basic information about the ownership, nature, roles, and capabilities of these crucial machines that orbit the earth.

Such a situation has even led media theorist Geert Lovink to suggest that it is necessary to think of the figure of the satellite in contemporary culture in psychoanalytical terms—as an unconscious apparatus that lurks behind the more obvious or "conscious" circuits of culture.[7]

"Publics around the world have both been excluded from and/or remained silent within important discussions about [the] ongoing development and use [of satellites]," Parks and Schwoch stress. "Since the uses of satellites have historically been so heavily militarized and corporatized, we need critical and artistic strategies that imagine and suggest ways of struggling over their meanings and uses."[8]

## Ultimate High Ground

Space superiority is not our birthright, but it is our destiny. . . . Space superiority is our day-to-day mission. Space supremacy is our vision for the future.[9]

Even a preliminary study of the world of satellites must conclude that they have contributed powerfully to the extreme globalization of the contemporary age. Nowhere is this more apparent than in the murky and clandestine worlds of military and security surveillance satellites. Not surprisingly, the idea of colonizing inner space with the best possible satellite sensors has long made military theorists drool. Their pronouncements revivify long-standing military assumptions that to be above, physically, is to be dominant and strategically in control of the subjugation of enemies.

Whistle-blowers and leaks only occasionally hint at the extraordinary powers of globe-spanning military and security satellites. By communicating details about the earth's surface—its geography and other attributes—to secretive ground stations and by allowing weapons like drones to be directed anywhere on earth from a single spot, military and security satellites produce what geographer Denis Cosgrove has termed "an altered spatiality of globalization."[10]

German theorist Peter Sloterdijk stresses the way the increasing dominance of the view of the earth from satellites, since the 1960s, has revolutionized human imagination about the planet through what he calls "inverted astronomy."

"The view from a satellite makes possible a Copernican revolution in outlook," he writes. "For all earlier human beings, gazing up to the heavens was akin to a naive preliminary stage of a philosophical thinking beyond this world and a spontaneous elevation towards contemplation of infinity. . . . Ever since the early 1960s an inverted astronomy has thus come into being, looking down from space onto the earth rather than from the ground up into the skies."[11]

This sense of global, total, and seemingly omniscient vision from above allows military satellite operators in particular to render everything on the earth's surface as an object and as a target, organized through near-instantaneous data transmission linking sensors to weapons systems.[12]

Crucially, such "virtual" visions of the world, wrapped up in their military techno-speak, acronyms, and euphemisms, are stripped of their inherent biases, selectivity, subjectivity, and limits. The ways in which these visions are used to actively and subjectively manufacture—rather than impassively "sense" the targets to be surveilled and, if necessary, destroyed—is consequently denied.

A further problem, of course, is that satellite imaging efforts also completely ignore the rights, views, and needs of those on the receiving end of the technology on the earth's surface, far below satellite orbits—the people most affected by the domineering technology above.

As with the closely allied worlds of the drone, military helicopter, and bomber, this imperial trick works powerfully. It images the world below as nothing but an infinite field of targets to be sensed and destroyed, remotely, on a whim, as deemed appropriate by operators in distant bunkers. "All the various aspects of satellite imagery systems . . . work together," writes communications scholar Chad Harris. They do this, he says, to create and maintain "an imperial subjectivity or 'gaze' that connects the visual with practices of global control."[13]

To deny how constructed and subjective satellite visioning is, militaries and security agencies present it as an entirely objective and omniscient means for a distant observer to represent the observed. The God-like view of satellite imagery is often invoked by states as evidence of unparalleled veracity and authenticity when they are alleged to depict, for example, weapons of mass destruction facilities, human rights abuses, or nefarious military activities.

It does not help that many critical theorists mistakenly suggest that contemporary spy satellites effectively have no technological limitations, or that a hundred Hollywood action movies—erroneously depicting spy satellites as capable of witnessing anything—do the same. All too often, critics depict

---

[7] Lisa Parks interviewed by Geert Lovink, "Out There: Exploring Satellite Awareness," *Institute of Network Cultures*, November 1, 2005, http://networkcultures.org/blog/2005/11/01/out-there-exploring-satellite-awareness/.

[8] Parks, "Orbital Performers."

[9] Gen. Lance W. Lord, then commander of the United States Air Force Space Command, cited in Tim Weiner, "Air Force Seeks Bush's Approval for Space Weapons Programs," *New York Times*, May 18, 2005, http://www.nytimes.com/2005/05/18/business/air-force-seeks-bushs-approval-for-space-weapons-programs.html.

[10] Denis Cosgrove, *Apollo's Eye: A Cartographic Genealogy of the Earth in the Western Imagination* (Baltimore: Johns Hopkins University Press, 2001), 236.

[11] Peter Sloterdijk, *Versprechen auf Deutsch* (Frankfurt: Surhkamp, 1990), as cited in Wolfgang Sachs, *Planet Dialectics: Explorations in Environment and Development* (Chicago: University of Chicago Press, 1999), 110–11.

[12] Philosopher Rey Chow calls this "the age of the world target." See Chow, *The Age of the World Target: Self-Referentiality in War, Theory, and Comparative Work* (Durham, NC: Duke University Press, 2006).

[13] Chad Harris, "The Omniscient Eye: Satellite Imagery, 'Battlespace Awareness,' and the Structures of the Imperial Gaze," *Surveillance & Society*, 4, no. 1/2 (2006): 101–22, http://www.surveillance-and-society.org/Articles4(1)/satellite.pdf.

satellite surveillance as totally omnipotent and omniscient, and our world subject to complete dystopian control with no limits to the transparency of the view, and no possibilities for resistance or contestation.[14]

In suggesting, for example, that "the orbital weapons [and satellites] currently in play possess the traditional attributes of the divine: omnivoyance and omnipresence,"[15] French theorist Paul Virilio radically underplays the limits, biases, and subjectivities that shape the targeting of the terrestrial surface by satellites.

Instead of invoking satellites as an absolute form of imperial vision, it is necessary, rather, to see satellite imaging as a highly biased form of visualizing, or even simulating, the earth's surface rather than some objective or apolitical transmission of its "truth."[16]

Where maps are now widely understood to contain partiality and even error, satellite images are still widely assumed to present a simple, direct, and truthful correlation of the earth. This occurs even when there is a long history of such images being so imperfect and uncertain—that is, manipulated, mislabeled, and just plain wrong—that it is necessary to be skeptical of such claims.[17]

## World-Zoom: Google Earth

Today the aerial view—the image of everywhere—seems to be everywhere.[18]

Perhaps the most profound effect of the contemporary proliferation of satellites centers on the way their extraordinary powers of seeing from above are now harnessed to personal computers and smartphones.

Google Earth is especially pivotal here. As a system of systems linked to a computer or mobile smartphone, it offers almost infinite possibilities for zooming into and out of views of the earth's surface at local, regional, and global scales.[19]

It does this by "mashing up" global satellite imagery, geopositioning coordinates, digital cartography, geolocation data, 3D computerized maps, architectural drawings, street-level digital imagery, and other social media, data, and software. These are configured together as an "always-on," interactive, and boundless datascape—a flexible and multi-scaled portal of primarily vertical images which now mediate life in new and important ways.

The apparently infinite "scale-jumping" possibilities of Google Earth force us to revisit, and update, a very long-standing debate about the politics of the aerial, "God's eye," or top-down view.[20] Many cultural theorists argue that the new ubiquity of the digital view from above is an important part of a contemporary shift away from a world dominated by a stable and single sense of ground and horizon organized through linear perspective. Instead, contemporary societies are saturated by a multitude of "always-on" digital and screen-based perspectives; extending armies of prosthetic eyes layered across entire volumes of geographic space; intense and real-time globalization; and, for many, unprecedented human mobility. Satellites and satellite vision are absolutely pivotal to the new sense of vertical free fall that attends this new age.[21]

Google Earth is pivotal to these transformations. It is the prime means through which vertical and oblique views of our world have very rapidly become radically accessible, zoomable, and pannable in myriad ways. Many researchers suggest that mass public access to Google Earth fundamentally challenges long-standing assumptions that the view from above necessarily involves dispassionate, technocratic, or privileged visual power.[22]

In presenting a "virtual globe" that can be navigated on screen and repeatedly zoomed, Google Earth introduces a powerful imagination of the planet—one that is simultaneously global, corporate, and saturated with commercial data and location-based advertising. It is thus "closely related to the production and movements of contemporary urbanization."[23]

The active shaping of this virtual globe by the viewer is crucial, however. In contrast to media like aerial or satellite photographs, users of Google Earth are not simply passive viewers witnessing the world in zoom.

Instead, users actively customize their experience of Google Earth by designing their own interfaces within the system, adding their own data and imagery.[24] Indeed, the frame-by-frame animation of the Google Earth interface works to provide users with a virtual globe they can manipulate to produce a personal cinematic rendition of the planet that can

---

14   See Paul Kingsbury and John Paul Jones, "Walter Benjamin's Dionysian Adventures on Google Earth," *Geoforum* 40, no. 4 (2009): 502–13.

15   Paul Virilio, *Desert Screen: War at the Speed of Light*, (2002; repr., New York: A&C Black, 2005), 53.

16   See John Pickles, ed., *Ground Truth: The Social Implications of Geographic Information Systems* (New York: Guilford Press, 1994).

17   The satellite images of purported WMD facilities presented by the US government to justify the 2003 invasion of Iraq are a sobering example here. See David Shim, "Seeing from Above: The Geopolitics of Satellite Vision and North Korea," in GIGA Institute of Asian Studies, *Working Paper 201* (Hamburg: German Institute for Global and Area Studies, 2012), http://giga.hamburg/de/publication/seeing-from-above-the-geopolitics-of-satellite-vision-and-north-korea.

18   Mark Dorrian and Frédéric Pousin, eds., *Seeing from Above: The Aerial View in Visual Culture* (London: I.B. Tauris, 2013), 295.

19   Ursula Heise, *Sense of Place and Sense of Planet: The Environmental Imagination of the Global* (Oxford: Oxford University Press, 2008): 11.

20   See Chris Tong, "Ecology without Scale: Unthinking the World Zoom," *Animation* 9, no. 2 (2014): 196–211.

21   See Hito Steyerl, "In Free Fall: A Thought Experiment on Vertical Perspective," *E-Flux Journal* 24 (April 2011), http://www.e-flux.com/journal/in-free-fall-a-thought-experiment-on-vertical-perspective/.

22   Mark Dorrian, "On Google Earth," in *New Geographies 4: Scales of the Earth*, ed. El Hadi Jazairy (Cambridge, MA: Harvard University Graduate School of Design, 2011): 164–70.

23   Daniel Laforest, "The Satellite, the Screen, and the City: On Google Earth and the Life Narrative," *International Journal of Cultural Studies* 19, no. 6 (July 2015): 659–72.

24   Tong, "Ecology without Scale."

then be viewed and traversed in a decidedly God-like way. Media scholar Leon Gurevitch calls this the "divine manufacture of the very [Google Earth] environments [viewers] wish to travel through."[25]

The addition of street-level visuals throughout Google Street View, however, grounds this virtual world with imagery of current and historical street scenes. Now cloud- and street-level worlds work seamlessly together, shimmering visual surfaces that occlude as much as they reveal to the inspecting subject. The system's interface "provides the ability to come and go freely within a completely controlled universe," media scholar Daniel Laforest emphasizes, "while maintaining the sense of distance as a constant promise, a source of leisure, or even as an unexpected pleasure."[26]

Despite its flexibility, the cultural and political biases of Google Earth are not hard to spot. Until recently, the system defaulted to a view that placed the US at the center of the screen. The interface offers little evidence of the source or accuracy of the global surveillance that sustains Google Earth. The way the system itself collects reams of data that is passed on to commercial information markets or security and surveillance services like the US Department of Defense's National Security Agency (NSA) is also carefully obscured.

In Google Earth, many areas of the globe are also censored or offered at deliberately low resolution. Under US law, Google must represent certain parts of Israel/Palestine at low resolution. States have also been found to doctor Google Earth images. Hawkish security commentators, who stress the usefulness of Google Earth to those planning terrorist attacks, are now urging that such censorship be extended. "Terrorists don't need to reconnoiter their target," Russian security official Lt. Gen. Leonid Sazhin said in 2005. "Now an American company is working for them."[27]

The social biases of Google Earth can also be stark. In post-Katrina New Orleans, for example, efforts to use Google Earth to allow communities affected by the crisis to share information and support across the various neighborhoods of the city were geared overwhelmingly toward more affluent and whiter neighborhoods because of wider geographies of the so-called digital divide in the city, governing which users could access the system and its data.[28]

Certain information, moreover, is dramatically prioritized within the system—information for users of corporate services, automobile drivers, and so on. Google Earth's dominant, de facto data sets are heavily dominated by a cluster of key transnational corporations. To sustain their competitive advantages in tourism, travel, leisure services, oil consumption, and food provision, these companies overlay Google Earth's satellite image surfaces with geolocation data geared toward exploiting the new screen interface.

Other information—say, documenting human rights abuses or revealing the installations of national security states—is obviously obscured or inaccessible, sometimes through the crudest censorship. Extreme biases in access and use, meanwhile, indicate that the user-generated content of Google Earth strongly reflects wider social and ethnic inequalities in society.

Beyond this lies a burgeoning politics of urban legibility and camouflage, as state, commercial, and non-state actors work to appropriate the new vertical views to conflicting ends. As financial collapse hit the Greek state in 2009, for example, the government tried to locate wealthy Athenians guilty of tax avoidance by using Google Earth to find their swimming pools. The immediate response was to drape tarpaulins over the telltale azure rectangles.

Meanwhile, many social and political movements have mobilized Google Earth and satellite imagery in their efforts to expose war crimes and state violence in places as diverse as Darfur, Zimbabwe, the Balkans, Syria, Burma, and Sri Lanka.[29] Satellite images have been very helpful in securing prosecutions against war criminals at the International Criminal Court in The Hague.[30]

Activists in Palestine, meanwhile, have actively used the system to generate maps that depict widening Israeli control there as an effort to undermine the cartography produced by the Israeli state to legitimize or minimize its degree of colonial control.[31] The system has also been a boon to those aiming to expose, hack into, or contest the scale and power of national security states, military forces, and corporate power.[32] Perhaps the most famous example here was the discovery in China in 2006 of a military training area that mimicked precisely the exact terrain of a section of the Indian-Chinese border that has been in dispute since 1962.

In Bahrain, in 2011, Google Earth's ability to trace aggressive efforts to vertically build up "reclaimed" land to fuel elite real estate speculation had a huge impact. The mass uprising of the Shiite majority against the dictatorial Sunni elite—brutally suppressed by local security forces with the help of Saudi paramilitaries—was ignited partly by the circulation of Google Earth images depicting the scale of corrupt land reclamation by those very elites, to radically remodel and further privatize the tiny nation's coastline.

Because Google Earth clearly has enormous potential to support activism and critique, it is easy to forget that such new, GPS-enabled activism fundamentally relies on "dual-use" devices that only function as the result of military rocket

---

25   Leon Gurevitch, "Google Warming: Google Earth as eco-machinima," *Convergence* 20, no. 1 (February 2014): 97.

26   See Laforest, "The Satellite, the Screen," 6.

27   Cited in Roger Stahl, "Becoming Bombs: 3D Animated Satellite Imagery and the Weaponization of the Civic Eye," *MediaTropes* 2, no. 2 (2010): 66.

28   Michael Crutcher and Matthew Zook, "Placemarks and Waterlines: Racialized Cyberscapes in Post-Katrina Google Earth," *Geoforum* 40, no. 4 (2009): 523–34.

29   See Lisa Parks, "Digging into Google Earth: An Analysis of 'Crisis in Darfur,'" *Geoforum* 40, no. 4 (2009): 535–45; Andrew Herscher, "From Target to Witness: Architecture, Satellite Surveillance, Human Rights," in *Architecture and Violence*, ed. Bechir Kenzari (Barcelona: Actar, 2010), 127–48.

30   James Walker, "Archimedean Witness: The Application of Remote Sensing as an Aid to Human Rights Prosecutions" (master's thesis, UCLA, 2014).

31   Linda Quiquivix, "Art of War, Art of Resistance: Palestinian Counter-Cartography on Google Earth," *Annals of the Association of American Geographers* 104, no. 3 (2014): 444–59.

32   Martin Dodge and Chris Perkins, "Satellite Imagery and the Spectacle of Secret Spaces," *Geoforum* 40, no. 4 (2009): 546–60.

launches. The system supporting such efforts also enables the deployment of a series of 24 geosynchronous satellites that continually drop deadly ordnance on a wide range of countries. It is a system inevitably mediated by imperial networks of militarized ground stations and data centers, that relies on a network of atomic clocks run by the US Air Force.

In such a context, media theorist Roger Stahl emphasizes the military origins of the whole aesthetic of Google Earth, which "began its life as the very picture of war." During the 2003 Iraq invasion, he relates,

> [a] certain 3D aesthetic appeared in the form of virtual flybys, as part of more complex computer animations, in studio surveys of bomb damage, in speculations on the whereabouts of Saddam Hussein, and a range of other uses. It is not an exaggeration to say that this aesthetic took center stage in the high-tech spectacle of US television coverage.[33]

Such a perspective forces a deep appreciation of the ways in which, despite its widening civilian use, Google Earth remains a highly militarized domain embedded fundamentally within a broader military-technology-geotechnology-security complex. This suggests that it is a key instrument through which citizens now consume state military violence, contributing to the mythology of "clean" war and "precision weapons" that the contemporary US military is eager to circulate. "Rather than say that the 3D satellite image has been 'demilitarized' as it has entered civilian life," Stahl emphasizes, "it may be more accurate to say that the transference has draped the planet with a militarized image of itself."

Finally, the vertical gaze of satellite imagery that is suddenly so remarkably accessible offers important new perspectives on how the horizontal geographies of our planet's surface are changing. For it is only from such distant heights that we can possibly begin to make sense of the extraordinary territorial formations currently being created by the rampant growth and sprawl of the world's urban areas.

"To truly exist," Rice University Professor of Architecture Lars Lerup writes, "every city needs its perspective. Its point of view. Its eyes."[34] And yet the dominant experience at the edges of many sprawling urban areas—beyond the clusters of rapidly rising skyscrapers and elite residential towers—is one of apparently endless horizontality. In such landscapes obtaining a sense of the wider city becomes all but impossible.

Google Earth allows such landscapes to be understood. Only the zoomable and extensive top-down gaze of the satellite can really stretch to encompass what Lerup calls the "striated, spread-out geographies" of contemporary urbanized regions and "megalopolitan" corridors.

Writing about Alan Berger's remarkable maps of the geographies of sprawl and wasted land in urban America, Lerup points out that "from a satellite, this neglected in-between (of drosscape or 'pure unadulterated waste') is the real grammar of the horizontal city, requiring a new mathematics whose nature, strength and intelligence lies embedded in its apparent incoherence."[35]

With the satellite view of the city now a normal way of representing urban areas for mass consumption, navigation, planning and, increasingly, marketing, it is perhaps the way in which cityscapes are engineered as "brandscapes"—visible from space—that may be the most immediate example of Google Earth's impact on the ground.

As Mark Dorrian writes, "The terrestrial surface itself becomes manipulated as a media surface, not just virtually on the Google Earth interface, but literally."[36] The democratization of verticality has important effects: in this new mass-market medium, corporations are now concerned with how their spaces and buildings look from satellites and aircraft.

In other cases—such as the demolition of a US Navy office complex that by coincidence resembled a swastika, when seen from above—built space has been reengineered due to unwanted vertical associations via Google Earth. More prosaically, of course, Google Earth is being used by a wide range of tourists and travelers as a new medium for anticipating and planning journeys and checking the validity of claims by the tourism industry, and by urban planners, architects, and clients in the production and design of urban development projects.

It is necessary to emphasize again the rapid emergence of megastructural urban landscapes that are carefully designed with their representation through Google Earth in mind. Most notable are the Palm Islands and The World archipelago in Dubai—gargantuan projects marketed as "today's great development epic."[37]

Here, civil engineering, land art, and landscape architecture blur together. They do so to create vast artificial islands designed as vehicles for real estate speculation with the prime marketing advantage of a unique appearance, via satellite, on the mobile Google Earth interfaces captured on a billion smartphones in a billion pockets and a billion laptops in a billion tote bags.

This text, adapted for this volume of *New Geographies*, was first published in the author's book *Vertical: The City From Satellites to Bunkers*, Verso, 2016.

---

33   Stahl, "Becoming Bombs," 67.

34   Lars Lerup, "Postscript: Vastlands Visited," in *Drosscape: Wasting Land in Urban America*, ed. Alan Berger (New York: Princeton Architectural Press, 2007), 242.

35   Lerup, "Vastlands Visited," 243.

36   Dorrian, "On Google Earth," 169.

37   Viorel Badescu and Richard B. Cathcart, *Macro-engineering Seawater in Unique Environments: Arid Lowlands and Water Bodies Rehabilitation* (London and New York: Springer, 2011), 63.

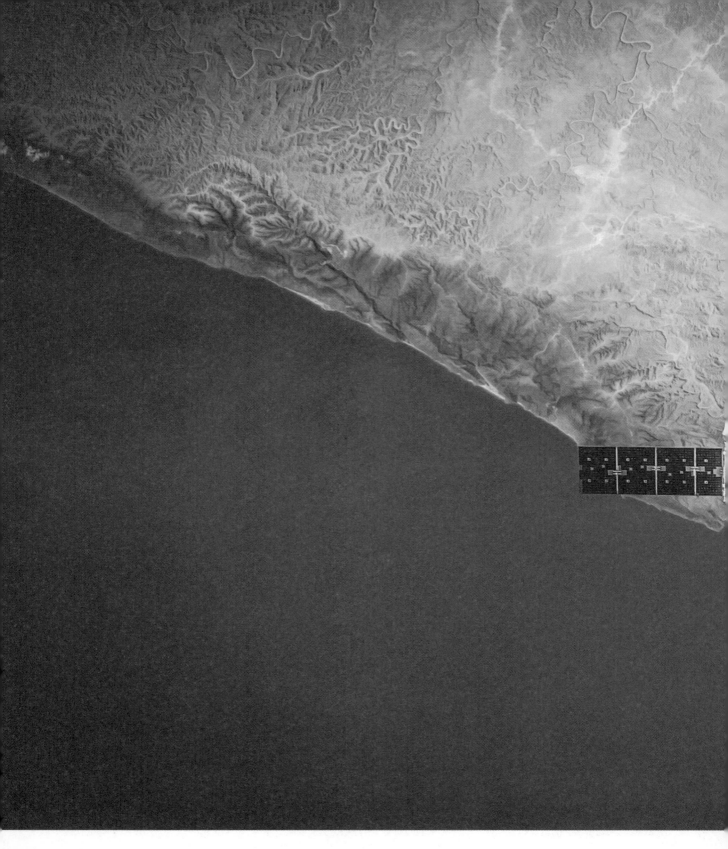

Satellite, 2016.

The techniques of policing and discipline and the choice between obedience and simulation that characterized the colonial and postcolonial potentate are gradually being replaced by an alternative that is more tragic because more extreme. Technologies of destruction have become more tactile, more anatomical and sensorial, in a context in which the choice is between life and death. — Achille Mbembe, 2003

# On the Alienated Violence of Money: Finance Capital, Value, and the Making of Monstrous Territories

**Martín Arboleda**

# Introduction

By 2005, just a few years before the worst financial meltdown in decades began its unrelenting crescendo of destruction, the (notional) value of outstanding financial derivatives was 630 trillion US dollars. This was equivalent to 14 times the global gross domestic product (GDP)—that is, 14 times the combined product of all national economies in the world.[1] With the cognitive maps that one regularly uses to go about the activities of everyday life, it is practically impossible to even begin to grasp the material life of that figure. Surely, one could picture how $1 million, or even $1 billion, might transform into things that can be touched, worn, and sensed. Beyond the threshold of $1 trillion, however, money begins to flee the worldly realm of human experience and becomes a mere artifact of mathematical abstraction. Reflecting on the nature of the infinite, Hegel posits a distinction between a genuine infinity and a "bad" or "spurious" infinity.[2] The question of boundlessness lies at the heart of these philosophical categories, because for Hegel, the self-identical object exists only by virtue of its confinement within a qualitative limit. A circle, the orbit of a planet, and a Möbius strip would be examples of genuinely infinite objects that have no outside and are thus complete in themselves. The truth of these objects, for that reason, is not external to their being.

Bad infinity [*die schlechte Unendlichkeit*], on the other hand, is linked to the notion of infinite progression. The straight line and the number series are regarded as canonical examples of the infinite, yet at any given moment they have a definite start and finish (any order of magnitude in the number series can be superseded by a larger number, and so on). For Hegel, bad infinity thus entails an unending process of linear progression and alteration that will never attain completion, precisely because such completion is logically impossible. When determined as a moment in the valorization of capital, the motion of money is, by its very essence, also without limit.[3] Money draws its immense material powers not only from the fact that it is formally boundless, but also due to its status as the universal representative of social wealth. As the social incarnation of all human labor, this perplexing super-object mediates an uncanny system of social and ecological transformation where the living, the inorganic, and the monstrous become entangled in the most improbable configurations. Rampant deforestation as lands are transformed into financial assets; suicide epidemics afflicting peasants under the burden of predatory lending; severe social anxiety resulting from mass evictions; and extermination of indigenous communities by death squads in order to make way for speculative investment are but a few of the imprints of money's ferocious logic of infinite progression on planetary natures.

The violent dislocations that emerge from the social powers of money in its character as store of value, however, are rarely problematized in the burgeoning literature on financialization. Most scholarly approaches have actually tended to place the focus almost exclusively on the function of money as a medium of circulation.[4] Emphasis is therefore given to the sphere of exchange, a domain of social reality mystified by the pervasive illusions of liberty, equality, and abstract citizenship that underpin the modern state-form. Such accounts thus tend to start from the explicit or tacit assumption that financial practices and instruments have become severed from the turbulent domain of "the real economy."[5] Yet, only by situating money in the sphere of production can we grasp its social determination as the universal bearer of human labor in the abstract, and therefore discover the origins of its overwhelming class power. Put differently, an exclusive focus on the credit system obfuscates the fact that labor exploitation and the appropriation of extra-human natures continue to be the pivot of the modern money-form. In rendering visible the clash between the bad infinity of money and the embodied realities of human and ecological existence, I intend to show that the process of financialization cannot be considered as separate—or even emancipated in the slightest—from the violent geographies that support it.

To formulate this point even more bluntly, the purpose of this article is to problematize the aesthetic and ideological disconnection that exists between the dazzling skylines of the financial centers of global cities, and the worlds of social suffering, genocidal war, and ecological destruction upon which they hinge. Even though the fetish of money gives the appearance that finance has become emancipated from production, neither of these polar opposites can exist without the other. In the first section I engage Marx's theory of money, and especially his appropriation of Hegel's notion of bad infinity, to make sense of the more-than-human powers of money in modern, liberal society. Then I historicize the specific context of global monetary space and its relation to the modalities of territorial change and industrial expansion that have come to define 21st-century capitalism. This involves the exploration of two key developments: the first is the emergence

---

1   Saskia Sassen, *Expulsions: Complexity and Brutality in the Global Economy* (Cambridge, MA: Harvard University Press, 2014).

2   The concept of infinity was introduced and developed by Hegel in *The Philosophy of Right*, *Science of Logic*, and the *Phenomenology of Spirit*. For more on the Marxian appropriation of this Hegelian notion, see David McNally, "Beyond the False Infinity of Capital: Dialectics and Self-Mediation in Marx's Theory of Freedom," in *New Dialectics and Political Economy*, ed. Robert Albritton and John Simoulidis (New York: Palgrave, 2003); Christopher J. Arthur, *The New Dialectic and Marx's Capital* (Leiden and Boston: Brill, 2002); Wayne Martin, "In Defense of Bad Infinity: A Fichtean Response to Hegel's *Differenzschrift*," *Hegel Bulletin* 28, no. 1/2 (2007): 168–87.

3   See Arthur, *The New Dialectic and Marx's Capital*; McNally, "Beyond the False Infinity of Capital."

4   Standard approaches in political economy ascribe a threefold social functionality to money: as a medium of exchange, money facilitates economic transactions and the trade of commodities; as a measure of value, money provides a standard of price that permits an assessment of the value of all other commodities; as a store of value, money is the general expression of wealth, and for this reason needs to bear an elementary connection to the values it represents. For a thorough explanation of the functions of money, see David Harvey, "Money, Credit, and Finance," in *The Limits to Capital* (1982; repr., London and New York: Verso, 2006), Chapter 9.

5   For a critique of traditional notions of financialization, see Susanne Soederberg, *Debtfare States and the Poverty Industry: Money, Discipline, and the Surplus Population* (London and New York: Routledge, 2014). On the embodied nature of finance, especially concerning the relation between financialization and the commodification of both labor-power and social reproduction, see Adrienne Roberts, "Gender, Financial Deepening, and the Production of Embodied Finance: Towards a Critical Feminist Analysis," *Global Society* 29, no. 1 (2015): 107–27.

of the floating currency system following the collapse of the Bretton Woods framework in 1971; the second is the aggressive monetization of socioecological processes that began to unfold in the aftermath of the 2008 financial crisis. After reflecting on the haunting forces and monstrous territories that have been unleashed by new, financially driven phases of primitive accumulation,[6] I conclude by arguing for the need to put Marxian monetary theory at the center of contemporary efforts to understand global sociospatial change.

## The Material Powers of Money

Hegel's influence on the work of Marx is usually considered only in relation to the early, more "philosophical" texts, such as *The Paris Manuscripts* and *The Poverty of Philosophy*. A recent strand of scholarship, however, has sought to reveal the methodological relevance that Hegelian thought exerted on Marx's mature intellectual production, and especially on his critique of political economy.[7] Marx's notion of capital as a formally boundless and self-expanding movement of valorization, in particular, starts from a definitive Hegelian basis, especially with respect to notions of limits and the elemental principle of self-movement that is immanent in forms. Hegel's organismic conception of nature led him to view reality as a living, interrelated, and self-actualizing whole. This aspect was formative in the work of Marx, who later framed modern society through the prism of a vitalistic materialism, where the world is seen not only as a dynamic system, but also imbued with self-generating powers.[8] In Volume 1 of *Capital* Marx compares the C-M-C circuit (selling in order to buy) typical of pre-modern/non-capitalist forms of social reproduction with the M-C-M' circuit (buying in order to sell) characteristic of capitalist modernity.[9] Referring to the latter circuit, Marx argues, "The end and the beginning are the same, money or exchange-value, and this very fact makes the movement an endless one." And he continues, "The circulation of money as capital is an end in itself, for the valorization of value takes place only within this constantly renewed movement. The movement of capital is therefore limitless."[10]

It is precisely due to this inversion of means and ends that the quantitative logic of money assumes almost fantastical attributes and begins to overturn all natural and social relations as it subsumes them into its restless, unbounded expansion. Money is a very peculiar commodity because, according to Marx, it is the universal representative of material wealth, and therefore the archetypal manifestation of value.[11] For this reason, money has the unique capacity to function as a universal equivalent (or store of value) and can therefore be used to put a price tag on practically anything: from water, to human virtue, to political loyalties, to the very air we breathe. Individuals, human groups, and even nation-states then become distinguished by the quantities of money they own. Indeed, as McNally suggests, the irruption of the money nexus into social relations has traditionally ushered in frightening and disorienting confusions between persons and things, as "money becomes animated with powers of life and death, and persons increasingly sell themselves, as if they were things."[12] In this confounding of human and nonhuman attributes, individuals therefore become mere personifications of the boundless, self-moving substance that is money. The capitalist "sacrifices the lusts of his flesh to the fetish of gold,"[13] while workers are compelled into debt and must sell their labor-power to avoid starvation and ill health. If the capitalist becomes crippled by the possession of money, the worker is devastated by a lack of it.[14] Not surprisingly, the poor throughout modern history have experienced the monetization of everyday life by means of a corporeal phenomenology of occult, fiendish, and diabolical forces that are beyond human control.[15]

Most importantly, reducing the embodied richness of socionatural relations to the impersonal abstractions of mathematical quantification, as David Graeber's forceful treatise on the history of debt and money shows, has invariably resulted in the crudest forms of violence.[16] The implications of his claim will rarely—if ever—be grasped if one examines only the sphere of exchange and its shopping malls, financial districts, and digital trading platforms. In its character as a commodity, money's value (i.e., creditworthiness) needs to be constantly preserved. This means that the life of money in modern society, whether in the form of bullion money sanctioned by state authority, or privately issued credit, cannot

---

6    I deliberately use the concept of primitive accumulation in order to signify a particular mode of social labor that is not part of a historical past from which capitalist social relations emerged, but as a form of labor premised on a logic of violent expropriation, which has continued to exist throughout the *longue durée* of modern history. For contemporary readings of primitive accumulation, see Werner Bonefeld, "Primitive Accumulation and Capitalist Accumulation: Notes on Social Constitution and Expropriation," *Science & Society* 75, no. 3 (2001): 379–99; Nikhil Pal Singh, "On Race, Violence, and So-Called Primitive Accumulation," *Social Text* 34, no. 3 (2016): 27–50.

7    Gastón Caligaris and Guido Starosta, "La crítica marxiana de la dialéctica hegeliana: de la reproducción ideal de un proceso ideal a la reproducción ideal de un proceso real," *Praxis Filosófica* 41 (2015): 81–112; Fred Moseley and Tony Smith, eds., *Marx's Capital and Hegel's Logic: A Reexamination* (Chicago: Haymarket, 2014); Moishe Postone, "Lukács and the Dialectical Critique of Capitalism" in *New Dialectics and Political Economy*, eds.

Robert Albritton and John Simoulidis (New York: Palgrave, 2003); Fred Moseley, *Money and Totality: A Macro-Monetary Interpretation of Marx's Logic in Capital and the End of the "Transformation Problem"* (Chicago: Haymarket Books, 2015); Arthur, *The New Dialectic*.

8    In particular, Hegel's monist metaphysics and the idea of the absolute as organism were of particular relevance in the rubric of materialism that Marx would later develop. See Helmut Reichelt, "Social Reality as Appearance: Some Notes on Marx's Conception of Reality" in *Human Dignity: Social Autonomy and the Critique of Capitalism*, eds. Werner Bonefeld and Kosmas Psychopedis (Aldershot: Ashgate, 2005).

9    C corresponds to "commodity"; M to "money"; and M' to "money plus profit" (surplus value).

10   Karl Marx, *Capital: A Critique of Political Economy*, vol. 1 (1867; repr., New York: Penguin Books, 1976), 252–53.

11   As cited in Arthur, *The New Dialectic*, 144.

12   David McNally, *Monsters of the Market: Zombies, Vampires, and Global Capitalism* (Chicago: Haymarket, 2011), 150.

13   Marx, *Capital*, vol 1, 231.

14   Terry Eagleton, *The Ideology of the Aesthetic* (Oxford: Blackwell, 1990).

15   See Michael Taussig, *The Devil and Commodity Fetishism in South America* (1980; repr., Chapel Hill: University of North Carolina Press, 2010); David McNally, "The Blood of the Commonwealth: War, the State, and the Making of World Money," *Historical Materialism* 22, no. 2 (2014): 3–32; Mary Louise Pratt, "Globalización, desmodernización, y el retorno de los monstruos," *Revista Histórica* 156 (2007): 13–29; McNally, *Monsters of the Market*.

16   David Graeber, *Debt: The First 5,000 Years* (2011; repr., Brooklyn and London: Melville House, 2014).

be disentangled from the practice of violence. Loan sharks send hired goons to collect at gunpoint; banks evict homeowners defaulting on their mortgage obligations; states erect vast military infrastructures as a means to give credibility to their currency. In fact, the history of modern central banking was the product of war finance. Starting with 16th-century Italian city-states, and later more systematically when the Bank of England began to fund colonial ventures in the 18th century, state money was baptized as war money.[17] This monetary paradigm, Graeber argues, has reached its most advanced expression with the cosmic power of the US military apparatus—and its singular capacity to drop bombs on any site on the planet in a matter of a few hours—acting as the pivot for the global economic system organized around the US dollar.

Insisting on the nature of money as the universal representation of abstract wealth, then, renders a completely different picture of the geographies of finance, prompting us to ask different questions. The very process by which an individual is integrated into the money nexus as either a "debtor" or "creditor" is far from mundane, as it usually entails a vast network of institutional mediations and material practices that frequently involve enclosures, expropriations, and the privatization of social goods. The very existence of the bourgeois state, it should be noted, springs from the necessity to generate a whole institutional materiality that not only grants specific groups the rights to issue credit moneys (via tax exemptions, concessions, bankruptcy laws, and other legal mechanisms), but that also incorporates the moral sentiments of enterprise and the price-form into the social fabric (via ideologies of home ownership, "financial inclusion" entrepreneurial behavior, etc.).[18] It is then by looking beyond the credit system that one can visualize how the magical properties of money (either in the form of cash, debt instruments, or speculative investment) systematically transform peasants into debt peons, people into living currency, and forests into wastelands. The aim of the next sections is to historicize and also develop a spatialized reading of these theoretical insights on the money-form. We begin this task through an interrogation of the monetary space that emerged after the abandonment of the Bretton Woods agreement in the 1970s.

## After 1971: Explosion of Money, Explosion of Spaces

The significance of the US administration's decision to terminate the convertibility of the US dollar to gold in 1971—thus ending the pegged exchange rate mechanism established as the foundation of the Bretton Woods system of postwar capitalism—can hardly be overestimated. During the last two millennia, most currencies were underpinned by a more or less direct link with precious metals. The move from bullion to virtual moneys, therefore, unleashed a wide array of new actors, practices, and sophisticated financial instruments that reconfigured the governance composition of transnational corporations and states. Of course, none of this seems particularly new if one takes seriously Giovanni Arrighi's claim that high finance has been present throughout the *longue durée* of the modern world economy, perhaps ever since Genoese bankers began to issue elaborate debt instruments in the long 16th century.[19] The scale, speed, and complexity of the post-Bretton Woods monetary apparatus, however, are completely unprecedented in human history.

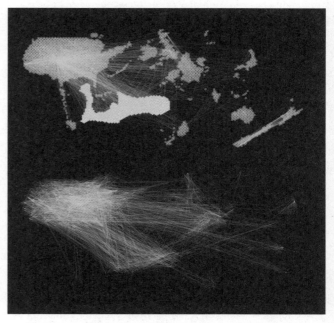

Visual representation of the Goldman Sachs corporate network, which includes thousands of subsidiaries registered in dozens of countries. Each dot represents a subsidiary, and each cluster of dots represents a country.

To put this in perspective, the daily turnover in foreign exchange markets amounted to $15 billion in 1973, and only 12 years later skyrocketed to $150 billion, soaring to $1.1 trillion by 1995. In 2007, daily turnover in foreign exchange trading had surpassed $3.2 trillion. Enabled by the technical prowess of computerization and by new modes of algorithmic expertise, new financial instruments such as derivatives sparked a revolutionary transformation in trading that quickly eclipsed stock and bond markets. In 2006, for example, $450 trillion in derivatives contracts were traded, a figure that dwarfs the $40 trillion global market for stocks in the same year.[20] New financial actors such as hedge funds, investment banks, and credit rating agencies gained

---

17  Graeber, *Debt*; McNally, "The Blood of the Commonwealth,"; Giovanni Arrighi, *The Long Twentieth Century: Money, Power, and the Origins of Our Time* (1994; repr., London and New York: Verso, 2010).

18  For the relation between money and the state, see Werner Bonefeld, "Adam Smith and Ordoliberalism: On the Political Form of Market Liberty," *Review of International Studies* 39 (2013): 233–50; McNally, "The Blood of the Commonwealth"; Simon Clarke, "The Rationality and Irrationality of Money," in *Value and the World Economy Today: Production, Finance, and Globalization*, eds. Richard Westra and Alan Zuege (New York: Palgrave McMillan).

19  Arrighi, *The Long Twentieth Century*.

20  McNally, *Monsters of the Market*, 159–61.

prominence, while existing actors grew in breadth and organizational complexity. As scholarly discussions on the emergence of the "global city" have illustrated, an entirely new built environment was produced to support the entrepreneurial, cultural, and organizational milieus of these new masters of finance.[21] Cities like London, Frankfurt, Hong Kong, New York, and Tokyo became nodal points in the sprawling network of monetary and financial flows that gave distinctiveness to the geocultural *zeitgeist* of our time: "globalization." This, however, is only one side of the coin.

As Neil Smith noted, these virtual moneys began to put price tags on most (if not all) extra-human natures, from land to water, carbon dioxide, metals, and biodiversity, among others.[22] The marketization and subsequent banking of ecological commodities became a highly profitable activity, just as primary commodity production was increasingly subordinated to the logic of finance. Yet, whereas conventional narratives tend to consider these processes of financialization as somehow "free-floating" or emancipated from the extraction of surplus value from the labor process, a closer look at the evolution of the mining and oil industries reveals a completely different story. The financialization of the oil industry, as Mazen Labban shows, did not replace the key antagonism between labor and capital with a tension between corporate managers and financial markets.[23] In their race to please increasingly voracious and unyielding institutional investors, corporate managers adopted an investment strategy that led to an intensification of labor in tandem with material growth, and hence to the extension of the discipline of capital over living labor.[24] Mechanization and territorial expansion were aggressively pursued alongside cost-reduction strategies that included offshoring, outsourcing, streamlining, and layoffs.[25]

These disciplined investments—disciplined by the logic of finance, that is—soon transfigured into poisoned rivers, vast deforested areas, and ancient glaciers ripped to shreds by machineries of extraction. They also spawned actual deaths, personal indebtedness, and laboring bodies in pain.[26] Such human suffering was not only an outcome of the hardships wrought by wage repression and labor casualization, but also of the manifold injuries and fatalities resulting from poor (or negligent) occupational safety conditions for those under precarious work contracts.[27] Due to the insidiousness of these new financial dynamics, Anna Tsing argues that an entirely different landscape developed throughout the wildly expanding resource frontiers of the post-1971 era. What once were quiet scenes of forests, fields, and houses "had become wild terrains of danger, urgency, and destruction."[28] A similar development ensued in the food industry, where the unprecedented expansion of agro-industrial enclaves yielded an export glut resulting in a 39 percent decline in world food prices between 1975 and 1989.[29] This boom, Jason W. Moore explains, was driven by access to cheap credit and enabled producers to circumvent declining levels of productivity resulting from soil exhaustion, suboptimal technological innovation, and water scarcity. A new international system of debt peonage compelled debtor states in the so-called developing world to liberalize their agrarian sectors in order to generate foreign exchange and service their sovereign debts.[30]

The debt-driven agro-industrial regime that transformed the Global South into a world farm not only caused further toxification and deforestation, but also involved systematic expropriation and displacement of peasantries—very often through militarized violence and ruthless brute force. This generalized sociospatial push factor, aptly termed "global depeasantization" by Farshad Araghi, explains to a considerable extent the rampant sprawling of shantytowns and rapid dissolution of agrarian modes of existence that characterize the era of planetary urbanization.[31] None of these violent ecologies and material practices of plunder, however, can be grasped if one remains tethered to the digitized circuits of financial trading that populate the wealthy quarters of the global city. A critical theory of value is therefore fundamental to understanding the status of money as a commodity, and thus the origins of seemingly ethereal financial instruments in the bloodshed and alienated labor of a swelling world proletariat. To fully appreciate the implications of this contradiction, we now turn to the analysis of the post-2008 context.

---

21   Saskia Sassen, *The Global City: New York, London, Tokyo* (Princeton: Princeton University Press, 2001); John Friedmann and Goetz Wolff, "World City Formation: An Agenda for Research and Action," *International Journal of Urban and Regional Research* 6, no. 3 (1982): 309–44; Carlos de Mattos, *Globalización y metamorfosis urbana en América Latina* (Quito: OLACCHI, 2010).

22   Neil Smith, "Nature as Accumulation Strategy," *Socialist Register* 43 (2007): 1–21.

23   Mazen Labban, "Against Shareholder Value: Accumulation in the Oil Industry and the Biopolitics of Labor under Finance," *Antipode* 46, no. 2 (2014): 477–96; Mazen Labban, "Oil in Parallax: Scarcity, Markets and the Financialization of Accumulation," *Geoforum* 41 (2010): 541–52.

24   For the case of the mining industry, see Martín Arboleda, "Financialization, Totality, and Planetary Urbanization in the Chilean Andes," *Geoforum* 67 (2015): 4–13; Julie Ann de los Reyes, "Mining Shareholder Value: Institutional Shareholders, Transnational Corporations, and the Geography of Gold Mining," *Geoforum*, published online ahead of print, 29, December, 2016.

25   Labban, "Against Shareholder Value."

26   See Arboleda, "Financialization, Totality, and Planetary Urbanization in the Chilean Andes."

27   See Labban, "Against Shareholder Value."

28   Anna Tsing, *Friction: An Ethnography of Global Connection* (Oxford: Princeton University Press, 2005), 67.

29   Jason W. Moore, *Capitalism in the Web of Life: Ecology and the Accumulation of Capital* (London and New York: Verso, 2015), 255.

30   See Moore, *Capitalism in the Web of Life*, chapter 10.

31   Farshad Araghi, "Global Depeasantization, 1945-1990," *The Sociological Quarterly* 36, no. 2 (1995): 337–68. In the field of urban studies, Teresa Caldeira has addressed some of the spatial dynamics of depeasantization through the notion of "peripheral urbanization." See Teresa Caldeira, "Peripheral Urbanization: Autoconstruction, Transversal Logics, and Politics in Cities of the Global South," *Environment and Planning D: Society and Space* 35, no. 1 (2017): 3–20. The idea of globalization as a problematic explosion of spaces was initially developed by Henri Lefebvre in *The Urban Revolution* (1970; repr., Minneapolis: University of Minnesota Press, 2003), and has been taken forward by scholars in the emerging field of planetary urbanization. See Neil Brenner, "The Hinterland, Urbanized?," *Architectural Design* (July/August 2016): 118–27.

# Post-Crash Geographies and Late-Capitalist Monstrosity

As the edifice of predatory financial surplus extraction engineered by Wall Street went tumbling down in 2008, it became clear that the credit system was far from emancipated from its monetary basis. As of 2009, write-downs on global assets were estimated at about $4 trillion.[32] Crudely put, this was virtual money that simply vanished because it was only functioning as a medium of circulation and had no substantive basis in the values it supposedly represented. The gears abruptly shifted to the realities of hard cash, opening the door onto a new phase of primitive accumulation whose full extent we are only beginning to understand. After the bust, a lack of confidence in financial markets obliged savvy financiers to reorient their investment strategies toward coarsely tangible assets: land, food, gold, and the relative surplus population. The bad infinity of money was set into motion yet again; this time its ties to labor exploitation and the appropriation of extra-human natures are too unambiguous to be taken for granted.

Saskia Sassen has noted that by 2006, bankers were already concerned about the impending financial catastrophe and had begun turning to assets with manifest "materiality" that could provide safe haven in the event of crisis. Land therefore surged as a key site for speculative investment, and it is estimated than 200 million hectares were acquired in the 2006–2011 period by venture capital firms. The graph in this page illustrates the post-2008 foreign direct investment spike that led to the new round of pillage, enclosures, and expropriations that has been referred to as "the new scramble for Africa."[33] Displaced peasantries, now part of the surplus populations that inhabit shantytowns and other precarious settlements, also became systematically targeted by new varieties of debt instruments. Microcredit emerged as one of the most lucrative industries after the financial collapse, especially when the G20 heralded it as a "core development strategy" for overcoming poverty and the global recessionary environment.[34] This class-based project has been draped in ideologies of "financial inclusion" so as to extend the discipline of monetary relations to the surplus population, thereby creating a global poverty industry that draws profits from 2.5 billion impoverished workers.[35] Moreover, land grabs and the resulting depopulation of the countryside made the post-2008 context ripe for a new phase in the financialization of agriculture. Investment banks and hedge funds became more directly involved in the agro-food supply chain, triggering price spikes, labor casualization, and food shortages.[36]

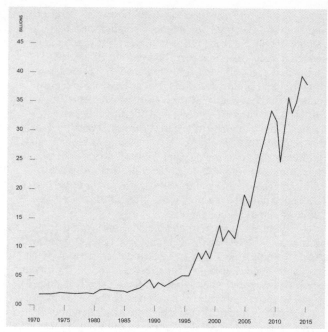

Net inflows of foreign direct investment in sub-Saharan Africa.

Having declared the abolition of magic and enchantment by modernity's project of unfettered rationalism, authors such as Max Weber and Max Horkheimer would surely be astonished to discover that the monetary relations of late capitalism have in fact signaled a return of the monstrous and the supernatural. Africa—the continent that has been most ravaged by the forces of globalization—is riddled with folktales spanning oral culture, viral videos, and pulp fiction that depict bewitched, devilish, and malevolent moneys. Zombies that spew money from their mouths, "bitter dollars" that destroy the lives of their possessors, and diamonds that devour people are among the local discourses that give popular meaning to the trenchant monetization of social reproduction that abounds in 21st-century sub-Saharan Africa.[37] Latin American vernacular culture also has its own host of monster tales signifying the violent irruption of the money-form upon the countryside. When the North American Free Trade Agreement (NAFTA) began to dismantle subsistence agriculture in the mid-1990s, peasants fell harshly into debt and stories of a monstrous vampire (*el chupacabras*) that slaughtered goats and attacked humans began to circulate in Mexico.[38]

Rumors of horrific creatures (*pishtakos*) in lab coats that would extract human fat for export to the United States—where it supposedly lubricated engines, computers,

---

32    "World Economic Outlook 2009: Executive Summary," International Monetary Fund, 2016, https://www.imf.org/external/pubs/ft/weo/2009/01/pdf/exesum.pdf.

33    Sassen, *Expulsions*, 80; Pádraig Carmody, *The New Scramble for Africa* (Malden, MA: Polity, 2013).

34    Soederberg, *Debtfare States*, 168; see also Philip McMichael, "The Land Grab and Corporate Food Regime Restructuring," *The Journal of Peasant Studies* 39, nos. 3–4 (2012): 681–701.

35    Soederberg, *Debtfare States*, 171; see also Katharine Rankin, "A Critical Geography of Poverty Finance," *Third World Quarterly* 34, no. 4 (2013): 547–68.

36    Frederick Kaufman, *Bet the Farm: How Food Stopped Being Food* (Hoboken, NJ: Wiley, 2013); Jennifer Clapp, "Financialization, Distance and Global Food Politics," *The Journal of Peasant Studies* 41, no. 5 (2014): 797–814.

37    McNally, *Monsters of the Market*; see also Jean Comaroff and John Comaroff, "Alien-Nation: Zombies, Immigrants, and Millennial Capitalism," *South Atlantic Quarterly* 101, no. 4 (2002): 779–805; Jean Comaroff and John Comaroff, "Occult Economies and the Violence of Abstraction: Notes from the South African Postcolony," *American Ethnologist* 26, no. 2 (1999): 279–303; Achille Mbembe, *Crítica de la razón negra: ensayo sobre el racismo contemporáneo* (Barcelona: Futuro Anterior Ediciones, 2016).

38    Pratt, "El retorno de los monstruos."

Martín Arboleda

and airplanes, and was used to pay the sovereign debts of Peru and Bolivia—also bred panic among communities in the rapidly urbanizing countryside.[39] In the US, the staggering growth of consumer debt that followed the 2008 crisis (estimated at roughly $11 trillion in 2015)[40] has also generated its own rubric of horror stories. The emergence of a gigantic industry that buys outstanding consumer debt from financial institutions, afterwards selling it to obscure collecting agencies, has given rise to uncanny forms of "zombie debt"—that is, debts that are charged despite being either nonexistent or already repaid to the original creditor. Strange telephone calls from anonymous, belligerent collectors threatening garnished wages, revoked driver's licenses, murdered pets, or kidnapped children are only some manifestations of these "undead" financial obligations.[41]

The encroachment and intensification of austerity measures that followed the 2011 financial crisis, coupled with the concomitant expansion of household and personal debt, have also severely disrupted the everyday fabric of European cityscapes as homelessness soars. In the United Kingdom, for example, the combination of a new wave of homeless populations and the proliferation of a highly toxic drug named "Spice" has transformed the squares of some city centers into nightmarish landscapes that, according to the media, seem directly out of a horror movie. Local workers, for example, point out how, in Manchester, "the sight of a person stumbling in a zombie-like state is as common in parts of the city center as someone selling coffee. . . . "[42] Mounting crowds of vulnerable young people yelling to walls, selling drugs, throwing up on park benches, or simply passing out in broad daylight, according to local media, have transformed some parts of Manchester into a veritable "hell on earth." The irruption of such horrifying devastation in the midst of the quotidian urban landscape leads a local health worker to say that it is as if there are two Manchesters now: "One where cool young urbanites find it buzzing with possibility and the other one, where people rot away in front of your eyes."[43]

The territorial embodiment of such multifarious modalities of economic and financial destitution is most patently manifested in the staggering, explosive growth of precarious, "informal" settlements across the planetary urban landscape. The resulting geographical configurations, which are often makeshift, irregular, and highly unequal, and geographically uneven environments (urban or otherwise), are part of a process that I propose to understand as *monstrous territorializations*. In this context, the monstrous city then

Industrial tailings pond.

emerges as the product and territorial domain of the excluded —in both geographical and ideological terms. Its monstrosity stems from the pervasive presence of the deformed, the wretched, and the anomalous (anomalous from the point of view of Eurocentric and neoliberal ideology, that is), expressed, for example, in precarious and highly heterodox architectural formations (autoconstruction, informal marketplaces, makeshift housing); economic activity (scavenging, street-hawking, hustling); and cultural practices (*patois* deformations of language, baroque aesthetics and rituals).[44] Monstrous sociospatial praxis, ensnared as it is in global monetary space, breeds cities within cities and wastelands beyond wastelands; monstrous territories that are at once internal and radically alien to a supposedly disenchanted modernity.

There is also something deeply monstrous and unsettling about the genre of aesthetic representation that falls under the label of geoaesthetics, or "the apocalyptic sublime," and which depict landscapes of environmental destruction and industrial debris that are so vast in scale as to transcend the capacity of human cognition. The frightening beauty of David Maisel's oblique aerial photographs of colossal tailings ponds filled with toxic effluent—which resemble the Abstract Expressionist paintings of Mark Rothko, another artist also

39   Ibid.

40   Holly Johnson, "The State of American Credit Card Debt in 2015," *The Simple Dollar*, June 3, 2015, http://www.thesimpledollar.com/the-state-of-american-credit-card-debt-in-2015/.

41   Suzanne McGee, "Zombie Debt Is Menacing America and Mine Even Has a Name: Kathryn, *The Guardian*, November 8, 2015, 2016, https://www.theguardian.com/money/us-money-blog/2015/nov/08/zombie-debt-is-menacing-america-and-mine-even-has-a-name-kathryn.; For scholarly intervention on zombies in global capitalism, see Chris Harman, *Zombie Capitalism: Global Crisis and the Relevance of Marx* (Chicago: Haymarket Books, 2013); Japhy Wilson, "Anamorphosis of Capital: Black Holes, Gothic Monsters, and the Will of God," in *Psychoanalysis and the Global*, ed. Ilan Kapoor (Lincoln: University of Nebraska Press, forthcoming).

42   Jennifer Williams, "The Pale, Wasted Figures Caught in a Spice Nightmare That's Turning Piccadilly Gardens into Hell on Earth," *Manchester Evening News*, April, 2017, http://www.manchester-eveningnews.co.uk/news/greater-manchester-news/spice-nightmare-manchester-city-centre-12870520.

43   Williams, "The Pale, Wasted Figures Caught in a Spice Nightmare That's Turning Piccadilly Gardens into Hell on Earth."

44   Verónica Gago, *La razón neoliberal: Economías barrocas y pragmática popular* (Madrid: Traficantes de Sueños, 2015); Michael Taussig, *My Cocaine Museum* (Chicago: University of Chicago Press, 2004); McNally, *Monsters of the Market*, chapter 3; Martín Arboleda and Daniel Banoub, "Market Monstrosity in Industrial Fishing: Capital as Subject and the Urbanization of Nature", *Social & Cultural Geography*, published online ahead of print December 17, 2016; Michael McIntyre and Heidi J. Nast, "Bio(necro)polis: Marx, Surplus Populations and the Spatial Dialectics of Reproduction and Race," *Antipode* 43, no. 5 (2011): 1465–88; see also Caldeira, "Peripheral Urbanization: Autoconstruction, Transversal Logics, and Politics in Cities of the Global South."

fascinated by the magical forces and dislocations of modern life—has something that is eerily reminiscent of the alienated abstractions that shape such landscapes. In fact, aesthetic theory has much to say about the ways in which the unbounded expanse of money resembles the archetypal figures of the sublime: the raging ocean, the looming electrical storm, the immensity of a starry night.[45] For Marx, Kantian aesthetics grasped the true essence of the sublime in the idea of the formless or monstrous [*Die Unform*].[46] The infinite movement of capitalist money (M-C-M'), accurately represented in the image of the spiral—the mathematical synthesis of a circular movement in the horizontal plane and a straight line in the vertical plane—[47] powerfully encapsulates the obliteration of forms at the heart of the sublime as an aesthetic category.

It is this spiraling, quasi-autonomous movement of money, seemingly beyond any limit or possible human control, which endlessly spawns the frightful creatures and damaged netherworlds that have come to populate the fabric of social and ecological existence in the 21st century. Overturning the alienated violence of money demands, first and foremost, full understanding of its concrete social determinations. For this reason, any critical-theoretical project that aspires to analyze the geographies of financial capitalism must place the sphere of production as its point of departure. Insisting on the character of money as the universal bearer of value will prompt us to transcend the mystified domains of markets and of exchange and ask why is it that debt instruments such as derivatives and credit cards, to paraphrase Marx, come to this world "dripping blood from every pore" in the first place. Money is transhistorical; a puzzling cultural artifact possibly as old as the human species. However, it is in capitalist modernity that we witness the perverse inversion between means and ends that renders it an end in itself. Liberating money from the swirling, infernal rhythms imparted to it by capital, then, cannot be accomplished without first shattering the chains and fetters of the laboring poor, which secure the reproduction of its substance—that is, value.

**Acknowledgments**: This paper has greatly benefited from conversations, comments, and constructive criticism by Neil Brenner, Japhy Wilson, Gediminas Lesutis, Mazen Labban, Alejandro Marambio-Tapia, as well as from the patience and careful editorial work of Ghazal Jafari and Mariano Gómez-Luque. I also thank Liat Racin for inviting me to present some of the ideas of this paper at her course on urban political ecology at Tufts University. Many thanks are also due to David Maisel and Kiln Digital for generously granting permission to reproduce their visual work. Research for this paper has been generously funded by the Urban Studies Foundation.

45   See, for example, Terry Eagleton, *The Ideology of the Aesthetic*; Antonio Negri, *Art & Multitude* (Malden, MA: Polity, 2011); Theodor W. Adorno, *Aesthetic Theory* (1970 ; repr., London and New York: Bloomsbury Revelations, 2013).

46   Eagleton, *The Ideology of the Aesthetic*, 213.

47   Arthur, *The New Dialectic*, 148.

**Image Credits**
97: Courtesy of Kiln. www.kiln.digital/projects/opencorporates.

99: Courtesy of World Bank. www.data.worldbank.org/indicator/BX.KLT.DINV.CD.WD?locations=ZG&view=chart.

100: Courtesy of David Maisel, from *Terminal Mirage* photo series.

Mine site, Poland, 2017.

In recent years, photographic work on colossal landscapes of industrialized resource extraction and environmental destruction—particularly in connection to the large-scale infrastructures required for the production of petrochemicals—has generated considerable attention both in the public sphere and among environmentalists, conservationists, landscape architects and geographers. In many of the most widely circulated images of such landscapes, the spectre of worldwide ecological destruction is depicted with such richly aestheticized abstraction that some commentators have described this genre using phrases such as the "toxic sublime" or the "apocalyptic sublime." — Neil Brenner, 2014

# Geopolitical Ecologies of Acceleration: The Human after Metal

**Mimi Sheller
with Esther Figueroa**

The 20th-century "Age of Aluminum" radically transformed the built techno-environments of design, architecture, transport, and industry, with continuing effects today.[1] Aluminum materialized crucial transnational processes in everyday life, while its energy-intensive production reshaped land use in massive global projects of modernization and development: hydropower plants, mines, company towns, air power, space travel, skyscrapers, satellite telecommunications, highways, homes, and suburbs. My work on the cultural and material history of aluminum, in collaboration with Jamaican documentary filmmaker Esther Figueroa, seeks to bring into view the political-ecological and geo-political connections between modern architecture, lightweight design, and the accelerated infrastructures of mobile modernity that span the extended built landscapes of planetary urbanization.[2]

Technologies of mobility embody (or capture) energy in their production; in their use (doing particular kinds of work); and in their particular form, as determined by infrastructural moorings predicated on unequal access, such as liquid petroleum or the metal chassis (or frame) of an automobile.[3] Likewise, technologies of dwelling embody energy in their construction process; in their use (today dependent on electrification for many functions); and in their relation to unevenly geographically distributed infrastructures such as power generation, or "splintered" access to roadways, private vehicles, and hydrocarbons.[4] Focusing on aluminum's transnational role in reshaping modern mobility, streamlining design, changing global environments, and affecting everyday life, I apply critical political ecology to reenvision Eurocentric architectural theory and practice in a global context of the circulating materials (and pollution) that enabled the architectural program, industrial landscapes, and infrastructure spaces of modernization.

Materialist approaches to the geoecologies and global political economies that shape sea, earth, and sky space bring into view the deeper time of infrastructures based on oil, carbon, ocean currents, and the mining of metals. My approach to infrastructure draws on theories of "infrastructuring" as an active practice, critical logistics studies, and critical mobilities theory.[5] It also encompasses emerging work on vertical geographies that account for the vertical axes of urban infrastructures—from deep-sea mining and undersea cables to skyscrapers and satellite towers, orbiting satellites, aerial atmospheres, outer space, and interstellar bodies.[6] This suggests the importance of understanding infrastructure as embedded in deep geoecologies that are composed of elements found underground, under the sea, in the air, and in space, as well as on land, inside buildings, and even inside our bodies.

Bohn Corporation Ad:
"Forecasting by Bohn," *TIME* magazine, 1945.

Aluminum, the world's most widely used nonferrous metal, is abundant, lightweight, durable, malleable, non-corrosive, and chemically nonreactive. It became one of the most important materials of 20th-century modernity, enabling innovation in the transmission of electricity; land, sea, and air transportation; explosives, munitions, and warfare; energy extraction and distribution; design, architecture, and construction; domestic appliances, mass consumption, and the movement of many of the goods and services that we now take for granted. The extraction of bauxite and the industrial applications of aluminum contribute to "worldmaking" and "worldbreaking" processes. There is a simultaneous making of modern infrastructures of mobility (associated with dreams of lightness, speed, and modernity) alongside the creation of transnational regimes of immobility, including the maldevelopment of tropical countries, the forced displacement of deeply rooted indigenous communities through the imposition of hydroelectric plants and mining, and the ecological destruction and health impacts of mining and smelting.

---

1   This essay draws on Mimi Sheller, *Aluminum Dreams: The Making of Light Modernity* (Cambridge, MA: MIT Press, 2014). It is currently being expanded through a documentary film, *Fly Me to the Moon*, with director Esther Figueroa, funded by the Graham Foundation for Advanced Research in the Fine Arts.

2   Neil Brenner and Christian Schmid, "Planetary Urbanization," in *Implosions/Explosions: Towards a Study of Planetary Urbanization*, ed. Neil Brenner (Berlin: Jovis, 2014).

3   Kevin Hannam, Mimi Sheller, John Urry, "Mobilities, Immobilities, and Moorings," *Mobilities* 1, no. 1 (March 2006): 1–22.

4   Stephen Graham and Simon Marvin, *Splintering Urbanism: Networked Infrastructure, Technological Mobilities, and the Urban Condition* (London: Routledge, 2001).

5   Susan Leigh Star, "The Ethnography of Infrastructure," *American Behavioral Scientist* 43, no. 3 (1999): 377–91; Deborah Cowen, *The Deadly Life of Logistics: Mapping Violence in Global Trade* (Minneapolis: University of Minnesota Press, 2014); Keller Easterling, *Extrastatecraft: The Power of Infrastructure Space* (London: Verso, 2016).

6   Stephen Graham, *Vertical* (London: Verso, 2016).

"Rush Hour in Space," *TIME* magazine cover, June 1960.

I seek to explore the creation of an aluminum-based world that values high speed, lightness, and streamlined modernity, yet builds this dream of modernization by polluting the ecologies, land, and health of places understood as non-modern—a relation shaped by the transnational dynamics of political economy, uneven development, and ecological externalities. Aluminum production also introduced new landscapes—open-pit bauxite mines, lakes of toxic red mud, atmospheres transformed by air pollution, and vast stretches of flooded land—that resulted in displaced communities across the Caribbean, indigenous North America, India, Africa, and Australia. The "externalities" that the industry, its engineers, and its historians have largely ignored—pollution, energy use, and health impacts on workers, consumers, and those displaced by large-scale industrial development—are all crucial to understanding aluminum's ambiguous cultural history, as well as its contested future. Aluminum contributed to the stark contrast between modernity and backwardness, speed and slowness, mobility and immobility, as culturally representative of the differences between the "developed" and "developing" world.

Aluminum is also embedded in our imagination of what it means to be modern: we inhabit a world of skyscrapers, airplanes, space exploration, fast cars, fast food, and information that is available at the touch of a smartphone or click of a keyboard. In what ways has aluminum contributed to our perception of particular places as centers of modernity and others as peripheries of extraction? How can we combine perspectives on the political economy of commodity chains with cultural approaches that link sites of production and consumption? How does the use of aluminum for electricity, improved mobility, and architecture epitomize cosmopolitan life and the accelerated speed of global connections? How can we analyze zones of extraction as sites of intense social struggles and environmental damage? And what stories can we tell by foregrounding the materiality of aluminum in our attempts to comprehend the modern world?

The discovery of industrial-scale aluminum production processes in the late 19th century unlocked a new material culture of mobility and unleashed a technological drive toward acceleration in speed and lightness. The physical qualities of this light metal and its alloys contributed to the existing dream of high-speed travel and gravity-defying flight by finally making it possible. The electrochemical smelting of aluminum (discovered simultaneously in 1886 by two 23-year-old inventors, Paul Héroult in France and Charles Martin Hall in the USA) opened whole new vistas in the quest for speed and lightness. The smelting process brought within reach the apotheosis of speed, new architectures of luminosity, and the conquest of air and outer space that 19th-century writers like Jules Verne had thrilled at, and even feared. The dreams realized by aluminum became definitive of 20th-century modernity and its visions of the future. Aluminum became crucial to the making of modernity not only as a new material out of which to make particular objects (especially those that we associate with streamlined modernism), but also as a means of innovating across the entire infrastructure of transport and communication (underlying many of the technologies that we associate with modernization).

American philosopher, architect, and inventor R. Buckminster Fuller called for a technological transition that he named "emergence by emergency." He believed that the radical transition needed to solve global problems would only happen through the demilitarization of territorial control and a more equitable use of energy. Climate change and the imperative to reduce greenhouse gases may be precisely such an emergency. In 1933, Fuller designed the Dymaxion Car, a three-wheeled, streamlined aluminum-skinned vehicle that could carry up to 10 passengers. And his automated, lightweight, and mobile Dymaxion House, from 1927, was designed to be mass-produced, affordable, easily transportable (packed down into a single metal tube), and environmentally efficient. Fuller later invented other living systems including the Growth House (1952), which was based on similar principles to the Dymaxion House, but also featured space for an indoor, vertically hanging food-growing system. Even more spectacular were Fuller's urban designs, including Old Man River's City (1971), a proposal to contain the entire city of East St. Louis, Illinois, in a crater-shaped structure with 500-foot-high walls, sheltered under a one-mile-in-diameter "geodesic-sky parasol-umbrella" made of aluminum and stainless-steel trussing; and the tetrahedronal floating city that he was commissioned to design for Tokyo Bay (Tetrahedron City Project, 1968).[7]

---

7   R. Buckminster Fuller and Kiyoshi Kuromiya, *Critical Path* (New York: St. Martin's Press, 1982): 315–23, 333.

Aluminum innovations would indeed play a central role in the "grand domestic revolution," but as social historian Dolores Hayden shows, it was not exactly the radical revolution sought by feminists and other promoters of collective housing and futuristic lifestyles.[8] Instead of shared kitchens, food delivery services, and communal nurseries, modernity would foster the individualized, privatized suburban dwelling in which wives were isolated from public life and men drove automobiles to distant workplaces. The rise of the food-packaging industry, as well as domestic appliances such as freezers, were crucial to this process, which had deep impacts on the scheduling of daily life.[9]

The ease of high-speed travel, the convenient food packaging, the mobile communications systems, and the shining skyscrapers that serve today as the cathedrals of late modernity are grounded in the heavy (and dirty) industries of power generation, mining, refining, and smelting—and these are part of a worldwide production process controlled by a handful of huge multinational corporations with exotic acronyms: RUSAL, ALCOA, Alcan, CHINALCO, Rio Tinto, BHP Billiton, Vedanta. These corporate powers have been accused of ignoring the rights of local people, especially disempowered indigenous peoples in the tropics; of harming workers and nearby communities with dangerous chemical outputs; of consuming vast amounts of energy, which in turn contributes to global warming; and of exploiting the resources of poorer countries. Meanwhile, their owners accrue vast personal wealth.

Aluminum in its various forms and products is the final stage in an industrial process that begins with the mining of bauxite, the principal ore of aluminum. Bauxite is strip-mined, then processed into alumina, which is afterwards smelted into aluminum. The aluminum industry is intrinsically global, with bauxite extraction taking place, for example, in Australia, Guinea, Brazil, Jamaica, Guyana, Indonesia, Vietnam, and India; and smelting at production plants in Canada, Australia, the United States, China, Russia, Dubai, India, Iceland, and Norway. Understanding the history, contemporary developments, and future of the aluminum industry is crucial for understanding how truly globalized industries function, and how our most ordinary and everyday experiences are shaped by complex infrastructural spaces about which we are usually unaware.

For each ton of aluminum produced, four tons of bauxite must be washed, strained, baked, and dried into aluminum oxide, which creates four tons of caustic red mud-like residue. The oxide is shipped to smelt in places with cheap electricity, leaving behind red mud or red sludge as a waste product of the Bayer process which refines bauxite into aluminum oxide. Red mud cannot be disposed of easily, and yet 77 million tons are produced annually. It is pumped into holding ponds, after which, the surrounding land cannot be restored to farming. In October 2010, approximately one million cubic meters of red mud from an alumina plant in Hungary accidently spilled, killing ten people, destroying villages, and contaminating a large area that nearly spread into the Danube River.

Smelting requires 13,500 kWh of electricity per ton of aluminum, more energy than required to process any other metal. Smelting uses about 3 percent of energy worldwide, and averages 13 tons of $CO_2$ emissions per ton of aluminum. Smelters also account for 90 percent of tetrafluoromethane and 65 percent of hexafluoroethane emissions globally, both powerful greenhouse gases. Many industry facilities have turned to hydroelectric power, since it is considered a renewable clean energy source—preferable to burning coal or oil—and above all offers the round-the-clock, continuous current needed to keep the smelter pots from freezing solid. Yet in almost every case this has involved the displacement of indigenous and tribal peoples from their land and rivers, often accompanied by environmental destruction and human rights violations. From Niagara Falls and the Tennessee Valley River Authority in the 1920s to 1930s US, to Canada's Saguenay River projects and those along the Suriname River in South America by the 1960s, shifting to sites including the Zambezi River in Mozambique, the Three Gorges Dam in China, and tributaries of the Amazon in Brazil in the 1990s, "histories of aluminum and dam construction go hand in glove, linked from birth."[10] Dams that have been built primarily to supply the aluminum industry have flooded over 30,000 square kilometers of forested land worldwide. They have caused the relocation of over 200,000 local people from sites adjacent to the Nile in Egypt and the Caroni River in Venezuela, impinged on the territory of reindeer herds in Norway's fragile subalpine plateaus, destroyed habitat and threatened biodiversity in Brazilian and Asian rain forests, enabled the spread of debilitating tropical diseases in African valleys, and submerged archaeological treasures under water.[11]

More recently, huge new hydroelectric projects like the controversial Belo Monte Dam planned for the Xingu River, a tributary of the Amazon in Brazil, would send up to 25 percent of the total energy produced to nearby smelters owned by ALCOA (which is expanding operations across Amazonia) and Norwegian state-owned Norsk Hydro (which purchased assets from Vale do Rio Doce in 2010 including one of the world's largest bauxite mines at Paragominas, in northern Brazil). The Belo Monte project will displace remote "uncontacted" indigenous tribes, damage biodiversity, and add to greenhouse gas emissions as acres of flooded vegetation decompose, according to an International Rivers report (2010).[12]

As development has spread across remote aboriginal territories in Australia and untouched wilderness areas in Brazil, into tribal lands in Eastern India or politically unstable West African countries like Guinea or Sierra Leone, it is pertinent to ask who is responsible for monitoring and regulating transnational industries. The building of new aluminum

---

[8] Dolores Hayden, *The Grand Domestic Revolution: A History of Feminist Designs for American Homes, Neighborhoods, and Cities* (Cambridge, MA: MIT Press, 1982).

[9] Elizabeth Shove, *Comfort, Cleanliness and Convenience: The Social Organization of Normality* (New York: Berg, 2003).

[10] Felix Padel and Samarendra Das, *Out of This Earth: East Indian Adavasis and the Aluminium Cartel* (London: Orient Black Swan, 2010), 72.

[11] Jennifer Gitlitz, *Trashed Cans: The Global Environmental Impacts of Aluminum Can Wasting in America* (Arlington, VA: Container Recycling Institute, 2002), 24.

[12] International Rivers, "Belo Monte; Massive Dam Project Strikes at the Heart of the Amazon," *Belo Monte Fact Sheet* (March 2010). And see Juliet Pinto, Paola Prado, and J. Alejandro Tirado-Alcaraz, "Brazil and the Belo Monte Dam: 'The Amazon Is Ours'," in *Environmental News in South America*, eds. J. Pinto, P. Prado, and J. A. Tirado-Alcaraz (London: Palgrave, 2017): 81–114.

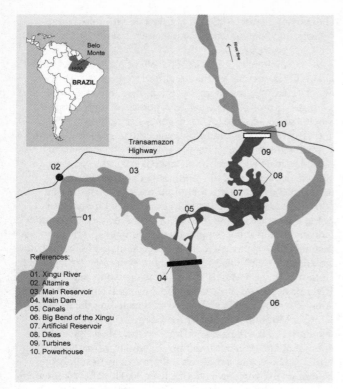

Map of Belo Monte Hydroelectric Project.

smelters and the hydroelectric dams that feed them is galvanizing opposition everywhere from Trinidad and Brazil to India and Iceland.[13] Even as global social movements unite in protest against the industry, it remains uncertain how environmental impacts can be minimized and how the completion of costly cleanups can be monitored, not to mention preventing harmful developments in the first place. In many countries the industry has been able to externalize environmental costs, leaving cleanup to future generations. Who will be able to influence corporate power and rein in a global industry with little accountability to any particular legal jurisdiction?

The tension between the desire for futuristic advanced technology and the desire to be earth-friendly is a sociotechnical dilemma that remains at the heart of our contemporary material culture and the drive toward innovation. It was never really resolved whether objects made from aluminum and other products of industrial processes are consistent with a "green" lifestyle or if they should in fact be banished altogether. Contemporary shifts toward corporate responsibility and movements for ethical consumption still play on this uncertainty, but they do not address the underlying global energy and resource (im)balance that informed Fuller's vision.

He called for the worldwide recycling of all metal scrap, and total reduction of waste.

Can societies afford the energy, pollution, and government subsidies that it takes to smelt aluminum? Is the switch from coal-powered electric generation to hydroelectric or geothermal power truly a sustainable solution? What are the associated ecological costs, and to what extent can we reduce demand not only for aluminum, but also for all kinds of resource extraction? Will moving into a post-petroleum and post-carbon economy also require that the world move into post-aluminum technologies, or do we need aluminum to create the sustainable technology solutions to global warming, and the spatiotemporal fixes for climate adaptation? What other forms of development might we imagine?

Neil Brenner and Christian Schmid's theory of planetary urbanization offers one approach that seeks to grasp these multiscalar mobilities of urban and planetary circulation. We need to pay more attention to the interactions between urban infrastructure, land use, and connectivity across "extended" urban systems and "operational landscapes" including mining, oil production, and water and energy flows, all of which feed into concentrated urban systems via transportation and logistics networks.[14] The long-distance global logistics chains that support extended urbanization are a planetary formation based on resource extraction and operational landscapes. As Martín Arboleda argues, these spaces of extraction are not only homogenized and concentrated urban spaces, but rather secured networks of flow that enable the intensification of extended urbanism to ever more distant places and its scaling up into ever more vast transregional agglomerations.[15]

Deb Cowen has recently noted the important role that infrastructure systems play in "profoundly material" crises, reminding us how "relations of power and of force rely on socio-technical systems that are themselves increasingly the object of struggle." Most importantly, Cowen urges us to consider how "infrastructure may entrench injustice in systems that seem technical rather than political, instead of technopolitical, and thus can serve to naturalize those relations. And infrastructure does not simply reflect existing inequality, but may engineer and entrench new forms."[16] Cowen's work on the militarization of logistics suggests the ways in which particular networks escape sovereign territoriality and legal jurisdiction (labor laws, environmental protection laws, etc.), are secured as militarized enclaves, providing safely connected corridors of mobility and communication.

At the same time, much of the primary demand for hydrocarbons and metals is driven by the military (i.e., for war, or war readiness), so the consumption side of the equation is also (in part) an artifact of geopolitics. And warfare is often concentrated on blowing up the infrastructure for oil

---

13  Jaap Krater and Miriam Rose, "Development of Iceland's Geothermal Energy Potential for Aluminum Production: A Critical Analysis," in *Sparking a Worldwide Energy Revolution: Social Struggles in the Transition to a Post-Petrol World*, ed. Kolya Abramsky (Edinburgh: AK Press, 2010), 319–33.

14  Neil Brenner and Christian Schmid, "Planetary Urbanization," in *Urban Constellations*, ed. Matthew Gandy (Berlin: Jovis, 2012), 10–13; Neil Brenner, ed. *Implosions/Explosions: Towards a Study of Planetary Urbanization* (Berlin: Jovis, 2014); Neil Brenner and Christian Schmid, "Towards a New Epistemology of the Urban?" *CITY* 19, nos. 2–3 (2015): 151–82.

15  Martín Arboleda, "Spaces of Extraction, Metropolitan Explosions: Planetary Urbanization and the Commodity Boom in Latin America," *International Journal of Urban and Regional Research* 40 no.1 (2016): 96–112.

16  Deborah Cowen, "Infrastructures of Empire and Resistance," *Verso Books Blog*, January 25, 2017, http://www.versobooks.com/blogs/3067-infrastructures-of-empire-and-resistance.

"Pare Belo Monte." Local resistance against the building of Belo Monte Dam in Brazil, in 2012, more than 23 years after the first Indigenous opposition against the construction of the dam in 1989.

production, to destroy supply chains. That means hydrocarbons and metals—the key commodities that determine the cost of many other goods—are valued to a large extent as a function of military power. Their provision and destruction effectively exist outside of market forces, yet the entire capitalist market economy depends on them since energy, machinery, and transportation (created with metals) are primary cost bases of many other commodities and products. Operational landscapes, therefore, are not simply artifacts of planetary capitalist urbanization, but also of state processes of coercive militarization, resource accumulation, and planetary logistics.

Planetary resource extraction has concentrated control over resources and energy not only in the hands of transnational private companies, but also in state hands, which has extended cross-regional and transnational infrastructure integration on a vast scale, outside the control of local democratic governance and beyond political transparency. Few advocates of technological transition fully explore how energy cultures are embedded in emerging technologies and systems that are crucial to military strategy, nor how everyday artifacts, infrastructures, and the routines they support arise from industries that are closely allied with military research and development, such as metal alloys research, which has direct military applications. This is why we need to excavate the deeper underground and above-earth histories of mining and metal production to understand the transnational material basis of contemporary post-ecological infrastructure spaces as posthuman built environments. This calls for what Caribbean theorist Sylvia Wynter names "the Human, after Man."[17] Being human must become a praxis that undoes the unjust infrastructure spaces of metallic modernity.

17   Sylvia Wynter, "Unsettling the Coloniality of Being/Power/Truth/Freedom: Towards the Human, after Man, Its Overrepresentation—An Argument," *New Centennial Review* 3, no. 3 (2003): 257–337; and see Katherine McKittrick, ed. *Sylvia Wynter: On Being Human as Praxis* (Durham, NC: Duke University Press, 2015).

**Image Credits**
105 & 106: Public Domain

108: Reproduction by editors based on data from Rio Times

109: Courtesy of Mario Tama from Getty Images

Hagadera, Kenya, refugee camp, 2016.

[E]xpulsions are made. The instruments for this making range from elementary policies to complex institutions, systems, and techniques that require specialized knowledge and intricate organizational formats. One example is the sharp rise in the complexity of financial instruments, the product of brilliant creative classes and advanced mathematics. . . . Another is the complexity of the legal and accounting features of the contracts enabling a sovereign government to acquire vast stretches of land in a foreign sovereign nation-state as a sort of extension of its own territory . . . Our advanced political economies have created a world where complexity too often tends to produce elementary brutalities. — Saskia Sassen, 2014

# Animal Life:
# A Visual Essay

# Jose Ahedo

### Introduction

While every discussion of livestock production today is inevitably linked to staggering numbers and statistical figures on energy use, the environment, and the production process itself, it is also true that little attention is paid to the design and living conditions of the people who work in these productive areas and live in the surrounding regions. Between 2014 and 2016, I traveled around 100,000 miles through some of these isolated rural territories on four continents. My intention was to observe and trace the domestic life, history, and culture that shaped those landscapes, to build a framework for speculation, and identify opportunities for design to have an impact in such a crucial sector.

### Mongolia, July 2015

A journey of over 3000-miles through very diverse ecosystems, from the Gobi Desert to the boreal forest along the border with Russia. Temperatures range from 104°F in the summer down to minus 40°F during the winter. The landscape is managed by discrete, family-based, semi-nomadic herding units. Domesticity is defined by a very high degree of adaptability to extreme environmental conditions. This adaptability, despite minimal resources, is achieved by the very efficient use of material systems and primitive technological solutions that work on multiple levels. A lattice serves both as an external structural element and an internal storage system to hang everyday life tools. The structural connectors are knots made from animal skin and lubricated with milkfat. Nothing is wasted; ropes are also made in the same way. Sheep felt provides a waterproof shell. Pastoral systems exemplify the recurrent technological search for a limit—a frontier beyond which architecture's control and domination dissolve into the wild.

### Azores, May 2015

São Miguel is the biggest, most populated island in the archipelago of the Azores. The economy based on orange production thrived until the late 19th century, when trade stopped and a plague of mycosis fungoides infected all the trees on the islands. A new economic reality based on dairy farming has since adjusted to the specific conditions of the unusual landscape. The island is an abrupt volcanic formation located in the middle of the Atlantic Ocean. São Miguel operates its dairy-farming industry today according to a rotational grazing system; farmers own several small, steep plots and animals are moved to graze from one location to another during the year. This constant movement between plots does not support the use of typical farm buildings. Rather, architectural artifacts—such as mobile milking parlors—help to resettle the farm functions wherever they are needed. These small pieces of machined architecture have enabled dairy farming in a nontraditional location, while maintaining the unique cultural identity of the island.

### United States, 1850s

Early American settlers established colonies on peninsulas and in other geographical nooks so that they could build the shortest possible fence to ensure their security and keep their livestock from straying.[1] Much later the settlement, encouraged by various Homestead Acts to expand the American frontier westward in the 19th century, was hindered by the lack of wood and stone in the Great Plains for building fences. Barbed wire hit the market around 1874.[2] This new technological invention was cheap to produce and easy to transport and install. The news of this architectural device spread quickly through country fairs and allowed vast pieces of land to be securely enclosed, which led to the conquest and domestication of territories previously unsuitable for European settlers.

### Bolivia, February 2016 and Paraguay, May 2016

No community embodies the tireless search for a new domestic frontier, or Promised Land, like the Mennonite community. The worldwide Mennonite diaspora dates to the Protestant Reformation of the16th century. Mennonite colonies have been continuously on the move, establishing new colonies and accepting offers from different governments to settle in remote, inhospitable lands where no other population wanted to work. Over time, these official state agreements specify the community's willingness to live in isolation, freedom to maintain religious traditions, educate their children in German, and not be obligated to perform military service or have any political affiliation. Despite the fact that many of the ceded territories encompass some of the most hostile environments and unproductive soils—from Alberta, Canada; to Monterrey, Mexico; Belize; Bolivia; and Gran Chaco—and that many Mennonites refuse to use electricity or any form of technology, the colonies have transformed into some of the most fertile and productive agricultural sites today. This change has been possible through a form of collective effort based on the sober devotion to work and a very austere lifestyle. The three main cooperatives established in El Chaco, Paraguay, during the 1930s—Neuland, Ferheim, and Chortitzer—have since engaged modern technology and adapted techniques for the region's salty soil and water. They still account for 75 percent of the nation's dairy production. During the late 20th century, both countries also engaged in other associative agricultural projects sponsored by the government, such as Nuclear Colonies in Choré (Paraguay) or the Low Lands project in Bolivia (also sponsored by the World Bank). While these architectural endeavors were meant to resettle and give free access to land for ethnic minorities (through an agrarian reform), the economic pressure of the soybean industry led to many of these lands being resold and acquired irregularly by national and Brazilian capital. They have also resulted in massive deforestation processes.

### Saudi Arabia, 2000s

The growth in demand for dairy products in the Middle East and the development of new technological processes have also facilitated the expansion of the domestic frontier to other unimaginable locations. The Saudi Arabian desert now accounts for some of the largest dairy farms in the world. Located in a region where temperatures can reach 122°F/50°C, some vertically integrated dairies can milk up to 75,000 cows daily. The scale of these companies and the need to secure animal feed has led many to acquire land on other continents—in places as far away as Sudan, Argentina, or California—to grow their own alfalfa and ship it 8,000 miles back to milking facilities. The traditional concept of a border frontier no longer applies; the countryside has been delocalized and the rural now depends on the transnational movement of commodities.

**New Zealand, August 2016**

Although New Zealand has no native mammals (except some species of bats), it is one of the largest producers of milk, beef, farmed venison, and wool in the world. With a population of 4.4 million people, it also accounts for 10 million cattle (6.5 million for dairy), 29.1 million sheep, and 0.9 million deer. All the animals, insects, and plants were slowly introduced from countries all over the world. Macrocarpa and Monterey pines were brought from California to provide wind shelter; sheep and cattle were brought from Europe, and deer from Canada. Over the last 150 years of ecological management, these plants and animals have replaced former wilderness areas, forming a new hyper-engineered and artificial domesticated ecosystem. Because of environmental adaptation, areas like the Canterbury Plains and Southland support less than 0.5 percent of native vegetation.

**End Note**

Traditional understandings of the dynamics between human and animal, rural and urban, or natural and artificial have been outdated, as these relations are continuously shifting. Architecture and technology are still at the heart of this ongoing transformation. Our human hunger for domestication and control will continue to play a major role in a near future, where the countryside will have to support the economic, environmental, and cultural impact of sustaining 10 billion people.

---

1   Virginia DeJohn Anderson, *Creatures of Empire: How Domestic Animals Transformed Early America* (Oxford: Oxford University Press, 2004).

2   Henry D. and Frances T. McCallum, *The Wire That Fenced the West* (Norman: University of Oklahoma Press, 1965).

Jose Ahedo

In order of appearance:

### India
Woman and goat in Rajhastan. Relations between humans and animals are shifting continuously. Any rural project should consider the new set of conditions and challenges that the countryside faces today, rather than reading a complex reality through an 18th-century pastoral perspective.

### New Zealand
Flock of sheep near Geraldine. Both animals and houses are protected from extreme climate conditions by hedges like the one visible in the background. The hedges can reach 65 feet (19.8 meters) in height. Their density cuts the effect of high winds and prevents animals from straying.

### New Zealand
South Island, Canterbury Plains. Hedges provide shelter for the houses and livestock from harsh climate conditions such as strong winds. They are not cultivated for ecological or environmental purposes but rather function as structural aerodynamic devices, or windbreaks. The hedges are made from Macrocarpa or Monterey pine, species that were imported (along with livestock) to create an "artificial" landscape that is one of the most productive farming areas in the world.

### Mongolia
Two women milking goats near Tsetserleg. As with a family-based economy, domestic life is ruled by the daily tasks associated with animal husbandry. The lack of advanced technology and extreme environmental conditions make milking by hand a significant part of everyday life for Mongolian nomadic herders.

### Mongolia
A household near Tsogtt-Ovoo. Households are scattered throughout the territory. Nomads move continuously with their animals to find water and better grazing areas. Some of these family units can herd up to one thousand animals.

### Azores
A three-slotted mobile milking machine. Because each farmer owns several noncontiguous plots of land on which animals graze cyclically, moving from one plot to another several times during the year, there are few traditional farm buildings on the islands. Instead, mobile milking parlors and associated infrastructure (fuel and water) move to a new location when the livestock finish grazing a particular area.

### China
Fishing village on Hainan Island. The village is an association of small family-based aquaculture, farming mostly tilapia and crab.

### Bolivia
Radial communities near San Julian. These radial patterns are part of an agricultural relocation project (Tierras Bajas Project) led by the government during the early 1990s. Farmers were resettled from the Altiplano into small colonies in order to cultivate a previously depopulated area. Each town center accounts for public services and generally contains a school, a church, and some sports facilities. All of the radial towns are arranged in a grid and connected by roadways.

### Bolivia
Industrial chicken farm.

### Paraguay
House on communal land near Maciel. Interior view of a house situated along the perimeter of a communal grazing area. The houses are built according to the plan of an indigenous typology known as Kuláta Jovái, with two opposing perforated facades that allow cooling through cross ventilation.

# Plant Life:
# The Practice of
# "Working Together"

# Rosetta S. Elkin

Big Sandreed plant.

The word *plant* emerges from two Latin origins: *planta*, to sprout; and *plantare*, to fix in place. One origin insinuates movement, the other stasis. In French, *plante* offers the same inconsistency, so that the word itself is caught between its meanings as a noun that indicates development, and a verb that implies permanence. This modest etymology suggests an entry into the central argument of my research: plants are objectified as a fixed form of knowledge, such that their aliveness is no longer a subject worthy of human speculation. The plant is wedged between worlds: aggressively weedy or strikingly ornamental, beneficial or useless, either despised or desired as it is naturalized, pacified, or capitalized. All ensuing engagement with plants necessarily chronicles the project of domestication whereby the human subject cannot resist being the active agent, endowed with power and armed with prediction, and more recently, as a pioneer of innovation. Each step—each accumulation of expertise—transforms plant life into a measure of human knowledge. As *plantare* achievements continue to dismember the organism as a whole, plants are recast as tools of science. Images of the plant-object are fixed in the human imagination, a world of forms.

If the *plantare* perspective accrues for botanical scholarship, it necessarily finds its way into spatial practices such as geography, urban design, and landscape architecture. The ensuing results lead to environmental alterations that treat plants as tools, specifications, and statistics, absent agency, movement, behavior, or fundamental biological activity. On the one hand, ecologists warn of increased invasive species that threaten the rich diversity of native plants. Such reports are isolated to the management of ecosystems, such that identifying and preventing invasions converges with the impudence of crop yield models and genetically modified seed stock. In this model, agricultural monocultures are economically sanctioned while intrusion into established forest dynamics are critiqued. Most significantly, governments are poised to conserve inherently dynamic relationships while corporations protect production. By virtue of accepting the lineage of ecological truisms that are a product of botanical speculation, landscape architecture is particularly implicated in such debates.

Consider instead the definition of *planta*—to sprout. This verb-status fittingly attends to plants by means of their activity, and existence, a mode of being that helps shed the perception of fixity. *Planta* reveals that a plant is a process, a swarm of activity, and a dynamic planetary force.[1] Thus modified, plant knowledge can be activated by virtue of aliveness, which resonates with the ambition of spatial practices that operate at ecological, regional, and continental scales. By paying close attention to the philosophy of Isabelle Stengers, it becomes possible to recover the relationship between plant and human life, so as to understand how novel spatial practices might emerge from a closer reading of plant life. In particular, mobilizing the Stengerian suggestion of "learning to work together" subverts the role of the human expert in order to reengage a shared context between plants and humans. The embrace of plant life as a vivid entity might yield a remarkable new theoretical scientific ontology.

## Indicators of Fixed Practices

The *plantare* perspective is so firmly embedded in modes of inquiry that it is challenging to delaminate the layers of history that define static procedures. A common narrative is offered through the critique of early expansionism, as exemplified by the volume of literature that helps expose the social and ecological impacts of early botanical speculation.[2] Here, plants

---

1   There is significant debate over the definition of "plant," which is a term most often used to describe green, photosynthesizing organisms. Biologically such organisms are referred to as eukaryotic photoautotrophs. The Oxford dictionary gives the following definition: A living organism of the kind exemplified by trees, shrubs, herbs, grasses, ferns, and mosses, typically growing in a permanent site, absorbing water and inorganic substances through its roots, and synthesizing nutrients in its leaves by photosynthesis using the green pigment chlorophyll. OED Online. http://www.oed.com/viewdictionaryentry/Entry/11125.

2   See, for instance, Lorraine Daston and Peter Galison, *Objectivity* (New York: Zone Books, 2007); Richard Grove, *Green Imperialism: Colonial Expansion, Tropical Island Edens, and the Origins of Environmentalism, 1600–1860* (Cambridge: Cambridge University Press, 1995); and Londa L. Schiebinger and Claudia Swan, *Colonial Botany: Science, Commerce, and Politics in the Early Modern World* (Philadelphia: University of Pennsylvania Press, 2005).

are portrayed as an instrument of colonial power, especially through patterns of global trade and the rise of horticultural imports. In the context of 18th-century botany, political power was routed through the enterprise of finding, identifying, and collecting plants mainly for the purposes of increasing the range of medicinal cures, maintaining yield for an expanding population, and profiting from an injection of exploitable raw materials. Such *plantare* procedures help pacify, fix, and label the plant as a form of human knowledge achieved through the objectification of plant life. For instance, the practice of counting stamens is the basis of our entire binomial nomenclature system, lending taxonomical authority to similar species based on shared features. This process of deliberately counting, ordering, and naming was brilliantly designed by Linnaeus in order to distinguish useful from useless attributes. At the time, taking account of the biological world seemed a feasible mission.[3] Binomial research yielded a resource, a system, a language of parts, and finally, scientification of the plant. This dissociative enumeration in parts rarely gets pieced back together again, as scientific classification limits knowledge to the accumulation of traits. Form is prioritized over formation, and the whole perceived as an aggregate of exploitable features.

The ramifications of early plant trade are superimposed on the procedures that subsequently sanction the manipulation or protection of plants. Plant knowledge now proliferates through the authority of institutionalized scientific disciplines. If the 18th century reduced the plant to its constituent parts in order to sustain an exploitable resource, then the 19th century delineated plant life to generate authority in the form of scientific expertise. All current engagement with plant life derives from the professionalization of the botanical sciences, according to the disciplinary specialization that advanced in the 19th century.[4] Today, common procedures and attitudes toward plant life are entrenched in the methods that are sanctioned by expertise and calculation, including carbon offsetting calculations, greening initiatives, afforestation, and other environmental do-goodisms that paradoxically proliferate at the expense of plant life. Such strategies engage the familiar protocols of designers, who tend to shrink at the authority (and impenetrability) of scientific discourses and therefore promote resolutions based on already established metrics. It is precisely this translation of trust, in science and in problem solving, that is at the heart of all formal aesthetic practices, and of the design professions in particular. As a consequence, the field of science tends to discourage the sensuous, articulate, and communicative subject, further constraining any possibility of interdisciplinary or collective practice. Even today, plant life is analyzed as a fixed backdrop against human and animal intentions. Whether through use or expertise, the project of pacifying plants strengthens human authority. One practice exploits the plant itself, the other, knowledge of the plant.

As adjacent disciplines such as forestry, geography, environmental engineering, and landscape architecture proceed, they rely on a fixed interpretation of plant life, inherited through botanical science, which forms a fragmented intellectual foundation. These adjacent disciplines, too, relay botanical knowledge as fixed—or *plantare*—continuing a tradition of reading the world of plants absent agency, movement, behavior, or fundamental biological activity. This is even more surprising given the extraordinary evolutionary success of terrestrial plant life, and the fact that humans are dependent on plants for most of our material needs. There is no alternative to a narrative of dominance achieved by human authority because any adjacent methodology is considered conjecture and often disregarded as unscientific.

This is not to imply that "all science is bad," or that all botanical evidence inevitably suppresses plant life. Rather, if the plant has become an artifact, tool, or index, it has done so in order to advance a particular form of scientific progress. What is problematic is the unchecked translation of fixed scientific procedures into strategies of spatial practice. The question then emerges as to how the plant can retain its aliveness in order to collaborate or participate in the research or design process. How does the plant make itself known? When the plant is released from a desiccated one-dimensionality, it transforms into a living, breathing collection of slowly dividing cells. It sheds divisive labels that preclude movement. It is no longer defined by human management techniques, resource dependency, or (especially unpleasant) productive landscapes. Thus conceived, the *planta* perspective appreciates the remarkable behaviors that are exclusive to plants, and resists the urge to make reductive comparisons, including the recent theoretical division between the human and nonhuman.

## The Stuff of Nonhuman Theory

The scientization of plant life resonates with conflicts inherent in the production of knowledge, as explicated by Bruno Latour and Isabelle Stengers. In a Latourian sense, the translation of expertise contributes significantly to a revisionist history that has powerfully decentered humans and rendered nonhuman subjects active instigators. In a Stengerian sense, translation not only redistributes expertise but also has the capacity to align with another kind of knowledge production, one that is collaborative and that asserts equal agency for nonhuman subjects. Knowledge production, then, pertains less to translation per se than it does to persuasion. Stengers describes

---

[3] The task of counting plants might have appeared possible when considering the relatively low diversity of plant life found in temperate Europe, but remains unexhausted when applied to the vast biological diversity of other global biomes, such as the tropics.

[4] Stengers describes how science is mobilized by the state and capitalist enterprise to accumulate power, arguing that professional consensus and the consolidation of expertise served to invalidate local or artisanal knowledge during the 19th century. See Isabelle Stengers, *Power and Invention: Situating Science* (Minneapolis: University of Minnesota Press, 1997): "Perhaps it is on the basis of what began in the 19th century, and not on what one calls the 'origins of modern science' that we should conceive of present day science" (116).

processes that filter objects through human ideas, theories, and categories so that they emerge redesigned, or configured into new arrangements. Stengers does not describe the agency of design in her scholarship, but her account of how typical procedures reinforce solutions alludes to the familiar protocols of designers described above. But rather than a pacified world rendered compliant by the extension of human knowledge—*plantare* as a fixed backdrop—Stengers insists on *planta* agency. In her terms, "working together" is a means to reconfigure the relationship between human and nonhuman.

Paying close attention to the aliveness of plants helps to emphasize other useless antithetical binaries that reduce the world to categories, objects, or measures. Plant life resists such diminutions, appearing in the gap between theory and practice in much the same way that the literature on reductionism has expressed the limits of treating the nonhuman as a passive recipient of human knowledge.[5] For instance, Bruno Latour's concept of a "politics of things" and Donna Haraway's notion of the "encounter" between species help to explain how the world "out there" is apprehended, how knowledge is authorized, and information is situated beyond the laboratory.[6] In Latourian terms, the translation of expertise implicates the scientist in the process of generating material, who therefore also contributes to the revisionist history that has decentered humans and rendered greater agency to the stuff of the nonhuman.[7] Here, the act of translation both redistributes expertise and aligns with forms of knowledge production that find affinity with such worldly phenomena. While distinctions are recast, Latour remains primarily focused on the twisted encounters of humans, which have been manipulated by virtue of a nonhuman protagonist. Objects are bundled together with creative neologisms, such that the key differentiation is made between the human subject and—seemingly everything else in the world. This antithetical binary of human and nonhuman, prevalent in the humanities and social sciences, merely draws attention to the imbrication of categorization in knowledge production, while the plant remains firmly planted in its place. I would like to emphasize what could happen if the plant were no longer considered a *thing* in that translation.

In advocating a practice that acknowledges the scientization of plant life, further categories such as the nonhuman are superfluous. Almost 20 years ago, Bruno Latour sketched out the risk of specialization with fair warning: "It may sound as if we, too, are marching along the same path, in a hurried flight from truth and reason, fragmenting into ever smaller pieces the categories that keep the human mind forever removed from the presence of reality."[8] To move past the "nonhuman" discourse, it is helpful to consider how the constant activity of plant life resonates with concerns inherent in organic form, including the difficulty of generating materials, producing biological evidence, and claiming authority. Stengers, by contrast, refrains from escalating the nonhuman in social and political theory, claiming that the negative prefix—*non*—does nothing to prevent the proliferation of superfluous expertise. Instead, the label furthers the tendency to think in binaries and categories that assimilate knowing and dominating.[9] Stengers contends that nonhumans must be contemplated as existent; otherwise, they endure as objects, which necessarily circles back to problems of knowledge production. In this case, plant life is also something other than the stuff of nonhuman social theory.

## "Working with" Plant Life

For Stengers, knowledge production is independent from *translation*—in the Latourian sense—and is defined by an ongoing act of persuasion between scientists. This persuasion is articulated *de facto* yet remains a highly refined opinion of an absolute and distinctly human order. Convincing and arguing with a motive for recognition summarizes the relationship between humans searching to stabilize the enormity of biological life. Stengers articulates such normative scientific persuasions as the filtering of organisms through human ideas, theories, and categories. Consequently, they emerge from the persuasion redesigned, flattened, or configured into new arrangements that serve more science. Instead, Stengers proposes that knowledge production occur by including the agency of the organism: "How to succeed in 'working together' where the event does not occur, where phenomena continue (and seem able to continue) to speak in many voices: where they refuse to be reinvented as univocal witnesses."[10] Stengers's suggestion of "working together" implies a modified relationship between human and nonhuman that accepts reciprocity. The event, in this case, is the evidence or proof that makes the subject compliant enough to become a tool of persuasion. Thus, rather than using scientific knowledge to objectify the world, Stengers advocates a research process that is informed (and not just acknowledged) by the researched. Further, she insists that the sciences are at their best when predictability and control are replaced by intellectual enjoyment.[11] Applied to plant organisms, "working together" forces us to abandon the idea that they are fixed entities enduring only for human advantage, and engages the potential of a collaboration between human and plant life.

---

5    In particular, the "nonhuman" turn is reliant on Latour's explanation of how microorganisms were not the cause, but the witness to epidemics. See Bruno Latour, *The Pasteurization of France*, trans. Alan Sheridan and John Law (Cambridge, MA: Harvard University Press, 1988).

6    Latour casts science as a collective project. See Bruno Latour, *Science in Action* (Philadelphia: Open University Press, 1987); Bruno Latour, *We Have Never Been Modern* (Cambridge, MA: Harvard University Press, 1993). Haraway's insight into the historically situated practices of human primacy highlights the inequality between human and animal species. See Donna Haraway, *When Species Meet* (Minneapolis: University of Minnesota Press, 2008).

7    Latour rejects the traditional objectivity presumed by scientists working in the field and explores the task of generating research materials. See Latour, *Science in Action*, 108–21.

8    Bruno Latour, *Pandora's Hope* (Cambridge, MA: Harvard University Press, 1999), 21.

9    Isabelle Stengers, "Including Nonhumans in Political Theory: Opening Pandora's Box?" in *Political Matter: Technoscience, Democracy, and Public Life*, eds. Bruce Braun and Sarah J. Whatmore (Minneapolis: University of Minnesota Press, 2010), 3–33.

10   Stengers, *Power and Invention*, 89.

11   In *Power and Invention*, Stengers uses *jouissance* in her native French, which refers to a physical and intellectual joy. The term is more theoretical than the typical translation, which replaces *jouissance* with mere "pleasure."

The *planta* perspective builds on Stengers's idea that by accepting research as a collaborative process we can "learn the humor offered to us by reliable and yet multivocal evidence."[12] In other words, plant life can be realigned with design practice such that the production of knowledge is shared and inclusive. Stengers's distinction between subject and object resonates with familiar critiques of scientific progress and the practices of forced conformity. Yet her speculations take this argument further, by proposing that it is not enough to itemize or merely call attention to the discrepancies, failures, and instruments that have made the nonhuman world silent.

In much of her work, including *Power and Invention* (1997) and *The Invention of Modern Science* (2000), Stengers argues for scientific expertise that exposes objectivity and privilege.[13] More precisely, she advocates for a science that extends beyond the boundaries of event, investigation, and technical agreement; for a science that is sanctioned by collective achievement. For Stengers, authority is concealed in the production of objects (rather than the activation of subjects), which necessarily excludes what is unobservable to humans. The result is a "psychological dramatization" that echoes the "geological, geographical, biological, and ecological processes that create spaces, model and drastically alter landscapes, thereby determining the migrations, competitions, or mutual amplifications between processes of growth, proliferations, slow erosions, and brutal disintegrations."[14] Her argument reveals the epistemological reductionism of scientific strategies that suggest the world of processes can be made to conform once they are studied and objectified. For example, the prospect that the drift from linear conformity through an isolated experiment can be applied to the scale of intervention reveals the incompatibility between spatial practice and knowledge production. That *processes create space* echoes these incompatibilities because epistemic access necessarily arrests process for study. If the plant is arrested to further the sciences, then it has done so with great success. But design professionals who claim to work with the "built environment" equally manipulate space using the materials of the "living environment," which is another reason why learning to work together is so relevant to landscape architecture. Rather than explain the adverse effects of hard science or explicate obvious human blunders, Stengers attempts to offer an alternative. Working together resonates with opportunity outside of academic scientific practice and is infused with a rare optimism for the uncertainties of science.

Stengers typically samples from physics and psychoanalysis to structure her arguments. When advocating for the need to work together, however, she draws from a particular case in the natural sciences—the intriguing work of Barbara McClintock, a pioneer of cytology and genetics.[15] McClintock's research in genetics concerned the study of corn, or maize (*Zea mays*), one of the earliest known examples of plant domestication. Stengers likely stumbled onto McClintock's scholarship through her own interest in genetic modification, but her case study nonetheless provides a tremendous example of working with plants. McClintock appealed to the unknown character of maize by engaging imaginatively with the living plant, allowing its adaptations to enter into her investigations. She tracked the singularity of maize down to a microscopic scale and found mobility in the genes that had previously been defined by stability and fixity.[16] Stengers describes this breakthrough in the relationship between the scientist and the subject as an "intervention": "Her great *jouissance* was the moment when a 'small detail' destroyed a grand idea, a superb generalization, when she knew that the corn had, if I can express it this way, 'intervened' between her and her ideas."[17] In the case of McClintock, compliance to a given scientific definition in relation to the world was abandoned for a novel concept, supported by her observations of mobility. McClintock's investigations coincided with the birth of molecular biology, which in turn coincided with the recognition she achieved mid-career when she became a member of the National Academy of Sciences in 1944, and was elected president of the Genetics Society a year later.[18] At the time, the concepts of regulation and control drove geneticists to describe stable strings of genetic code, later replaced by McClintock's famous "jumping gene," for which she garnered a Nobel Prize in 1983. McClintock found animation in genes by letting herself be part of the experiment: "When I was really working with them I wasn't outside, I was down there. I was part of the system. I was right down there with them."[19] On at least one occasion, incorporating irreducible, animated activity into the process of gathering materials and generating data produced a remarkable event in the history of ideas.

## The Living Environment

Current research into the role of plants in shaping global climate patterns speculates that the earth's environmental history is actually written in plant evolution.[20] This stunning new research positions plants firmly in the center of the dialogue on climate change using evidence from the fossil record. If ancient fossils are the key to how life on earth was formed, then plant organisms may be freed from the dusty basements of museums and given due recognition for their major role in planetary history. By using the past to decode the future, David Beerling argues that the plant must be reconsidered as an object of scientific study: "The argument is that we

---

12  Stengers, *Power and Invention*, 90.

13  Her main arguments hinge on Alfred North Whitehead's exploration of "dynamic" philosophy and Thomas Kuhn's essential description of "normal" science—both of which also attend to the growing divisions between disciplines and subdisciplines. See Alfred N. Whitehead, *Process and Reality* (New York: Free Press, 1978) and Thomas Kuhn, *The Structure of Scientific Revolutions* (Chicago: University of Chicago Press, 1962).

14  Stengers, *Power and Invention*, 55.

15  Evelyn Fox Keller, *A Feeling for the Organism: The Life and Work of Barbara McClintock* (San Francisco: W. H. Freeman, 1983).

16  Necia Parker-Gibson, "Profiles in Science for Science Librarians: Barbara McClintock, Seeing What Is Different," *Science & Technology Libraries* 32, no. 4 (2013): 315–29.

17  Stengers, *Power and Invention*, 111.

18  Evelyn Fox Keller, *A Feeling for the Organism*, 4.

19  Ibid., 117.

20  For an in-depth discussion of experimental paleobotany, see David Beerling, *The Emerald Planet: How Plants Changed Earth's History* (Oxford: Oxford University Press, 2007).

must marry these traditional elements of geology with a focus on plants as living organisms to mount a frontal attack on the citadels of received wisdom and orthodoxy and reach a deeper understating of Earth's history."[21] What if previous scientific histories are reformed or reprogrammed so that plants emerge as live, actionable, and conscious characters that have also shaped life on our planet? How would that affect the concept of the Anthropocene?

    The plant is the only earthly organism that connects the atmosphere to the territory, literally linking the ground and the sky. The plant is a dynamic actor in this exchange, transforming and adapting to human influences that elevate the achievements of the built environment over the adaptive qualities of the living environment. In contrast, practices such as those enacted by McClintock and emphasized by Stengers resist the highly mechanized prevailing theories that urbanize the living environment, helping us gain traction in another, more collaborative direction. Here, collaboration is understood as the act of working together. Both sanctioned models of scientific inquiry, and the viability of long-held ecological or conservationist traditions, with their preposterous contemporary greening strategies, are at stake in activating plant life today. Moreover, how landscape architects specify, eradicate, insist on, and superimpose plant life within their designs can be significantly retooled. Within the pressing realities of a planetary turn and a changing climate, it is time to incorporate the entire living plant organism into practice. This is not a plea to apply systems ecology to the scale of the plant, just as it is not enough to decide to acknowledge that plants are nonhumans. The crucial point is to learn how new types of practices can emerge in the reciprocal relations, or co-production between plants and humans.

    If the -ologies and -isms of spatial practice could be bound to a more collaborative form of botanical research, then the planted world might be appreciated in wholly new ways. This would entail a reading of plants whereby fragmentation is superseded by purpose and agency. I argue that plant life has heretofore been obscured because of the difficulty of actually evaluating animate behavior. And "working together" resists speaking in terms of what is fixed, recognizing that plant traits do not fit nicely into human criteria such as advanced slowness, chemical communication, and concealed rhizography. To deny that plants could be active participants is to deny attributes that are valued in the study of animal or human mobility. The presumption that plant life is sessile and devoid of intelligence overlooks the qualities of biological and territorial aliveness, and ignores the creative collaborations that take place between human and plant life. Stengers's scholarship redirects the attention given to nonhuman categorization and other useless binaries that continue to exclude and pacify plant life. Her work advocates for the resourceful inventions and passions that have decentered the human, in order to bind practice to a more collaborative form of discovery. We must urgently redress modern scientific techniques that rejected everything that could not be measured or made measurable. Can the plant be studied for its own sake, without furthering the dichotomies of living and nonliving, human and nonhuman?

21    Ibid., 3.

**Image Credits**
125: Image courtesy of Cattle Raisers Museum, Fort Worth, Texas.

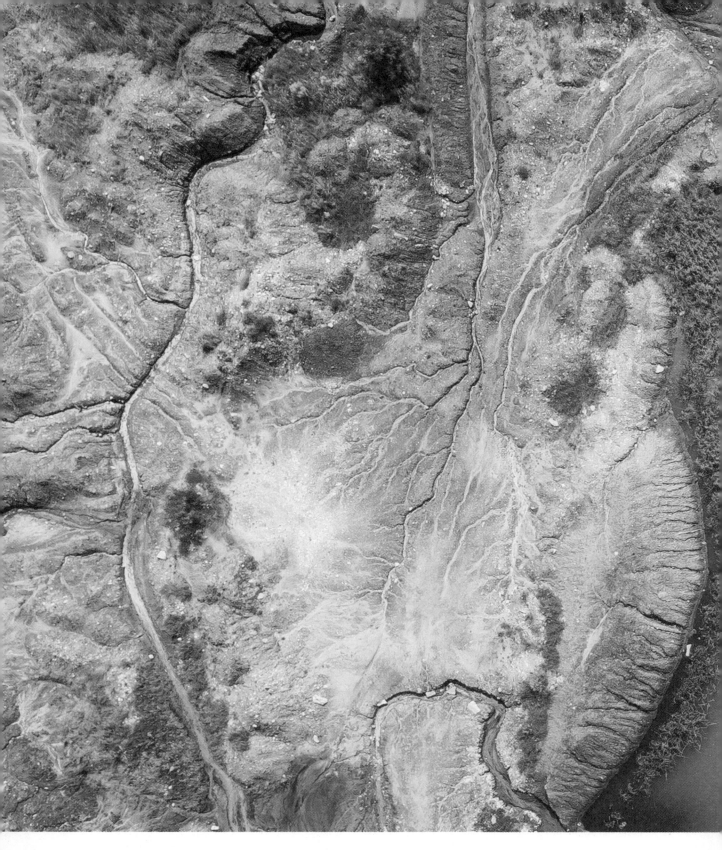
Old coal mine, Poland, 2016.

One downplays the suffering of plants by arguing that they have no central nervous system and thus cannot experience pain like animals can ... Plants sense the world, too, whether to seek out light or water, or to react chemically to external threats. — Ian Bogost, 2012

# "Walking to the Future in the Steps of Our Ancestors": Haudenosaunee Traditional Ecological Knowledge and Queer Time in the Climate Change Era

Eli Nelson

On July 3, 1991, Onondagan Turtle Clan Faithkeeper Oren Lyons met former White House press secretary Bill Moyers on Onondagan land in upstate New York for an interview on Moyers's public-access television program. Lyons, a representative of the Haudenosaunee—known also as the Six Nations, or Iroquois—attempted to communicate critical Haudenosaunee political concepts and insight, like the Great Law of Peace, and Six Nations traditional ecological knowledge (TEK). Yet he was repeatedly undercut as Moyers steered their conversation toward crime, drugs and alcohol, and violence in Haudenosaunee territory (and all across Indian Country) as definitive markers of American Indian modernity.[1] When Lyons turned the conversation back to the importance of TEK, Moyers exclaimed,

> More and more people are turning to Natives, to Indians and saying, "Share your knowledge with us." But don't you think honestly, now, practically—it is too late? For how many years have we had the dominant note, the clarion trumpet of "conquer the earth," you have dominion over the earth. We're building cities to create habitats far-removed from this. Our mentality is part of that "civilization" now and what you're talking about is wonderful and wise, the stories are profound and instructive and yet irrelevant—to this modern world.[2]

Unperturbed, Lyons replied, "It may be irrelevant at the moment. We don't preach here in this, our country, you know. We don't proselytize. . . . As a matter of fact, we try to protect what we have from intrusion."[3]

\*\*\*

In this paper, I trace the dynamics and political uses of Haudenosaunee TEK premodern temporality and Indigenous epistemic sovereignty, beginning in 1992 when TEK became an internationally recognized category following the United Nations Conference on Environment and Development (UNCED-Rio Earth Summit) and the international Indigenous Kari-Oca Conference. These meetings were prompted by mounting fears of anthropogenic climate change and threats to global biodiversity, both of which constitute "our ancestors' dystopia," as environmental philosopher Kyle Whyte has written, in that they indicate a severe depletion in cultural markers, nonhuman kin, and freedom of movement for Indigenous people.[4] UN representatives at the Rio Earth Summit feared this vision of dystopia for everyone else. In this context, TEK inhabited a queer discursive time-space, in which it acted as a "radical potentiality"—what José Muñoz in *Cruising Utopia* (2009) describes as "a thing that is present but not actually living in the present tense."[5]

At the Earth Summit, the Indigenous lack of present tense was due to a supposed premodern state, what seemed to them to be an enviable orientation as delegates struggled to imagine a future structured by a definition of humanity not set by the narrative of Western scientific progress. Theories of posthumanity abound in the climate change era. The Anthropocene, the proposed geological epoch marked by humanity's position as a prominent force in the earth's geosystems, has invited social scientific analyses of the end of humanity and nature alike.[6] In this posthuman and postnature view, landscapes have become more-than-human and less-than-nature, crafted by automated technologies and their wastes that now determine both human and natural history. This supposed faltering of the human-nature dichotomy has been mobilized to justify continued environmental extraction and dispossession in addition to calls to "stay with the trouble" and elevate interspecies collaboration.[7]

The Indigenous repudiation of present tense in the Kari-Oca Declaration, on the other hand, highlights the futurity inherent in radical potentialities. By refusing to assert the relevance of TEK in the modern world—echoing Oren Lyons's televised statement to Bill Moyers—the declaration, which Lyons helped to draft, queered Western epistemic sovereignty. As such, the temporal positioning and epistemic virtues of TEK provide a critique of posthuman models of the Anthropocene, even those that purport to undermine human-centeredness. In my conclusion, I will reflect on conceptions of posthuman geographies in the Anthropocene as seen from, and subverted by, the perspective of Haudenosaunee TEK.

---

1   TEK refers to a range of different Indigenous national and cultural bodies of knowledge and epistemologies that describe nonhuman and environmental relations, rooted in generations upon generations of Indigenous peoples living in and with such knowledge. Some commonalities between different varieties of TEK are a focus on systems thinking and interconnectivity, knowledge transmission that is oral and social, a lack of disciplinary boundary work, the inclusion of religious and moral insight, and environmental management practices that attempt to minimize the exhaustion of resources (including planting mutually beneficial crops, controlled burning, and migration). See Cassandra M. Brooks and Lisa T. Brooks, "The Reciprocity Principle and Traditional Ecological Knowledge: Understanding the Significance of Indigenous Protest on the Presumpscot River," *International Journal of Critical Indigenous Studies* 3, no. 2 (2010): 11–28; Lyn Carter, "Recovering Traditional Ecological Knowledge (TEK): Is It Always What It Seems?" *Transnational Curriculum Inquiry* 5, no. 1 (January 12, 2009): 16–25; Melissa K. Nelson, "Indigenous Science and Traditional Ecological Knowledge: Persistence in Place," in *The World of Indigenous North America*, ed. Robert Allen Warrior (New York: Routledge, 2015): 188–214.

2   Bill Moyers, "Oren Lyons the Faithkeeper," Moyers and Company, 57:05, July 3, 1991, http://billmoyers.com/content/oren-lyons-the-faithkeeper/.

3   Ibid.

4   Kyle Powys Whyte, "Our Ancestors' Dystopia: Indigenous Conservation and the Anthropocene," in *The Routledge Companion to the Environmental Humanities*, eds. Ursula K. Heise, Jon Christensen, and Michelle Niemann (London: Routledge, 2017).

5   José Esteban Muñoz, *Cruising Utopia: The Then and There of Queer Futurity* (New York: NYU Press, 2009), 9.

6   Dipesh Chakrabarty argues, for instance, that humanist boundaries between human and nature dissolve when the human figure becomes an integral component of nature as a global force, and natural history is then understood as synonymous with human history. See Chakrabarty, "The Climate of History: Four Theses," *Critical Inquiry* 35, no. 2 (January 1, 2009): 197–222.

Importantly, he also argues that anthropocenic subjectivities are at odds with renderings of the postcolonial subject in postcolonial theory. See Chakrabarty, "Postcolonial Studies and the Challenge of Climate Change," *New Literary History* 43, no. 1 (May 2012): 1–18. This kind of thinking was already circulating in the late 1980s in the work of environmentalists like Bill McKibben, who diagnosed the "end of nature" in humanity's expanded reach into a previously considered independent natural world. See McKibben, *The End of Nature* (New York: Random House, 1989).

7   Donna Haraway, *Staying with the Trouble: Making Kin in the Chthulucene* (Durham, NC: Duke University Press, 2016).

## TEK in Space-Time before 1992

The link between indigeneity and sound environmental relations was not new in the 1990s, particularly not in North America. The roots of the "ecological Indian" imaginary—the popular trope in settler discourse that envisions Indigenous land practices and religious traditions as inherently nondisruptive and harmonious—can be traced in part to turn-of-the-century shifts in conceptions of nature. Following economic and environmental crises in the first half of the 20th century, like the Dust Bowl and the 1929 US stock market crash, progressive reformers like John Collier, head of the Bureau of Indian Affairs in the administration of Franklin Roosevelt, viewed indigeneity as a balm for the ills of Western civilization; something to be preserved and circumscribed for posterity, much like the national park lands then being established.[8] That orientation soon became a staple of environmentalist discourse, as evident in the "Keep America Beautiful" campaign's infamous public service announcement from 1971, featuring Iron Eyes Cody as the "crying Indian chief."[9]

This is not to say that Indigenous people had not taken up the cause of protection of nonhumans against colonial abuses. The "ecological Indian" is a distortion of Indigenous practice, not a fabrication. Indigenous focus on environmental activism as a core political and ethical value was well established by the time those romantic tropes emerged. Indeed, the first protest (and later armed conflict) concerning a dam project in North America was staged in Boston by Polin, a Wabanaki leader. In 1739, he proclaimed: "I have to say something concerning the river which I belong to."[10] Haudenosaunee action on behalf of Kaniatarowanenneh—known also as the St. Lawrence River— followed a similar progression in the 20th century. Akwesasne Mohawk citizens protested when US and Canadian officials transformed the river into an industrial seaway in 1954, and the urgency of that work only increased over the next few decades as polychlorinated biphenyl (PCB) contamination of the river threatened nonhuman and human health alike.[11]

The international stage for these protests emerged in the mid-1970s, and Native nations and communities lobbied the United Nations for political recognition throughout the decade. Then newly established, the UN Working Group on Indigenous Populations—whose express purpose was to protect the fundamental freedoms of minority Indigenous populations in a postcolonial landscape—first met with Indigenous delegates in Geneva on July 5, 1982. Haudenosaunee representatives Oren Lyons and Cayuga Bear Clan Mother Carol Jacobs questioned the working group's priorities. Lyons later recalled:

> We went to Geneva—the Six Nations, the great Lakota nation—as representatives of Indigenous people of the Western Hemisphere, and what was the message that we gave? There is a hue and a cry for human rights—human rights, they said, for all people. And the Indigenous people said: what of the rights of the natural world? Where is the seat for the Buffalo or the Eagle? Who is representing them here in this forum? . . . We are Indigenous people to this land. We are like a conscience; we are small, but we are not a minority; we are the landholders, we are the land keepers; we are not a minority, for our brothers are the entire natural world and we are by far the majority.[12]

When Bill Moyers, in 1991, observed that more and more people were turning to Indigenous knowledge, he was right in theory. While there had been increasing awareness of TEK—or at least, a romanticized settler view of Indigenous knowledge—for years there had also been a great deal of friction. Native knowledge could not be extracted from Native people and politics, the survival of which unsettled the neocolonial landscape.

## Rio's Great Epistemic Break

A year later, the fragility of that landscape came into sharper focus. The agenda and principles developed at the 1992 Rio Earth Summit were the first produced by the UN to officially recognize the value of Indigenous "traditional knowledge and practices." And, unlike the 1972 Stockholm Conference, the Rio Earth Summit explicitly addressed climate change and expressed a deep anxiety about its ramifications.[13] The latter can be explained by the growing visibility and weight of climate change research in the 1980s and the founding of the Intergovernmental Panel on Climate Change (IPCC) in 1988, whose findings supported the reality of anthropogenic climate change and threats to global biodiversity.[14]

From June 3 to 14, 1992, the UNCED brought together 178 countries, representatives of 1,500 nongovernmental organizations, and a press corps of approximately 7,000 at the Rio de Janeiro convention center in Brazil to discuss the

---

8   John Collier, *On the Gleaming Way: Navajos, Eastern Pueblos, Zunis, Hopis, Apaches, and Their Land; and Their Meanings to the World* (Denver: Sage Books, 1962).

9   "The Crying Indian: 'Keep America Beautiful' Commercial," *NBC Nightly News*, New York: NBC Universal, October 15, 1978. For more on US settler conceptions of wilderness, nature, and the environment, see Roderick Nash, *Wilderness and the American Mind* (New Haven: Yale University Press, 2014). For more on US settler conceptions of Indigenous people, see Robert F. Berkhofer, *The White Man's Indian: Images of the American Indian from Columbus to the Present* (New York: Knopf, 1978).

10   Brooks and Brooks, "The Reciprocity Principle and Traditional Ecological Knowledge," 12.

11   Laurence M. Hauptman, *The Iroquois Struggle for Survival: World War II to Red Power* (Syracuse, NY: Syracuse University Press, 1986), 135; Elizabeth Hoover, "'We're Not Going to Be Guinea Pigs': Citizen Science and Environmental Health in a Native American Community," *Journal of Science Communication* 15, no. 1 (2016): A05.

12   Oren Lyons, "Our Mother Earth," in *Seeing God Everywhere: Essays on Nature and the Sacred* (Bloomington, IN: World Wisdom, Inc., 2003), 105.

13   Stanley Johnson and the United Nations Conference on Environment and Development (1992: Rio de Janeiro), *The Earth Summit: The United Nations Conference on Environment and Development (UNCED)* (London: Graham & Trotman/Martinus Nijhoff, 1993), 415.

14   Spencer R. Weart, *The Discovery of Global Warming* (Cambridge, MA: Harvard University Press, 2008), 173.

impending crisis. While the tone struck by the Rio Earth Summit was positively cheery in comparison to subsequent meetings on climate change and global environmental health, then UN secretary-general Boutros Boutros-Ghali did not shy away from the monumental shift
in discourse and self-understanding that climate change would necessitate. He boldly pronounced in his opening remarks:

> In the past, the individual was surrounded by nature so abundant that its immensity was terrifying.... All victories have been victories over nature, from the wild beasts menacing the cavemen to the distances separating communities.... Yet, "the time of the finite world" has come to an end.... There are no more oases to discover, no more "new frontiers." Progress, then, is not necessarily compatible with life.... It is this great epistemological break which the Earth Summit may ultimately symbolize for historians.[15]

Progress—as the ultimate political good, core tenet of truth-making, driver of empires, and author even of the UN—would need to be abandoned for more humble objectives. But the epistemological break Boutros-Ghali predicted never came to be; or, if anything, it was more of a fracture, the effects of which remain to be seen.[16] And yet, the provocation did invite imagined alternatives.

Later in his opening address, Boutros-Ghali characterized Indigenous people as "repositories of much of the traditional knowledge and wisdom from which modernization has separated most of us."[17] That is, TEK was framed as exclusively premodern—and as an innate feature of premodern humanity rather than something that is made or practiced. Produced at the close of the summit, Agenda 21 (the Rio Declaration on Environment and Development, and the Statement of Principles for the Sustainable Management of Forests) mentioned Indigenous peoples over 100 times, most often in lists of vulnerable populations to be protected. In the objectives of Chapter 26, which dealt exclusively with Indigenous peoples, Agenda 21 granted: "The lands of Indigenous people and their communities should be protected from activities that are environmentally unsound, or that Indigenous people concerned consider to be socially and culturally inappropriate."[18] Most of Chapter 26 covered legal recognition and protection for Indigenous peoples and the ways in which "sustainable development" could be achieved without violating the UN's commitment to Indigenous recognition. However, the fact that Indigenous peoples needed the same protection as the environment precluded them from Boutros-Ghali's vision of human history as the story of human domination over nature, and this was reflected in the document's treatment of Indigenous people as objects and resources of and for international policy.

## Native Space at Kari-Oca

In 1992, a journalist traveling with Haudenosaunee delegates to the Kari-Oca Earth Conference in Brazil was asked by Maria Pinta, a Mapuche woman, to inform the elders at Kari-Oca that the Native Mapuche people were still alive, despite Chilean dictator Augusto Pinochet's strategic announcement of their extinction in the 1980s.[19] In the 19th and 20th centuries, and now in the 21st, the non-native notion of Native cultures and peoples vanishing was ubiquitous. This vanishing race theory, rooted in social Darwinism, posits that Indigenous people will either be displaced by superior white societies or assimilate and therefore no longer be Indigenous.[20] Ongoing genocide and assimilation policies fuel that misconception, but it remains a categorical truism in the settler colonial mindset and its expression in the human sciences, which, in the words of anthropologist Johannes Fabian, takes Indigenous peoples as objects in the "there and then" to be known by the "here and now."[21] Indigenous people are denied history as they are frozen in a moment of pre-contact; classifying indigeneity as fragile and fleeting furthermore reinforces nonindigenous claims to land and authority.

Due to the prominence of the vanishing-Indian myth, when Indigenous communities demanded rights and recognition at decibels too loud to ignore, they could do so only as seemingly premodern subjects somehow emerging in late modernity. It is in this sense that TEK, as held by Native people, acted as a potentiality in the early 1990's. TEK was acknowledged as existing, but it could not exist in the present tense. José Muñoz writes:

> The queer futurity that I am describing is not an end but an opening or horizon. Queer utopia is a modality of critique that speaks to quotidian gestures as laden with potentiality. The queerness of queer futurity, like the blackness of a black radical tradition, is a relational and collective modality of endurance and support.[22]

Like queerness, indigeneity is a resource for making futurity, in that its (im)possible existence is predicated on, and constantly

---

[15] Johnson and UNCED, *The Earth Summit*, 55.

[16] Perhaps the most immediate evidence against such an epistemic break was the fact that the George H. W. Bush administration, although declaring the US a "leader, not a follower" in environmental protection, signed only a watered-down version of the Rio Agreement. My assessment is based in part on the use of "sustainable development" as a keyword at Rio, and the reformulated commitment to a universal progress narrative in "leapfrogging" discourses that also emerged in the early 1990s. For early examples of these discussions, see Ashoka Mody and Carl Dahlman, "Performance and Potential of Information Technology: An International Perspective," *World Development*

20, no. 12 (December 1992): 1703–19; Cristiano Antonelli, *The Diffusion of Advanced Telecommunications in Developing Countries* (Paris: OECD Publications and Information Center, 1991).

[17] Johnson and UNCED, *The Earth Summit*, 55.

[18] Agenda 21: Program of Action for Sustainable Development; Rio Declaration on Environment and Development; Statement of Forest Principles: The Final Text of Agreements Negotiated by Governments at the United Nations Conference on Environment and Development (UNCED), June 3–14, 1992, Rio de Janeiro, Brazil (New York: United Nations Dept. of Public Information, 1993).

[19] Elisabet Sahtouris, "Rio: The Indigenous Way," *Lifeweb*, 1992, https://ratical.org/LifeWeb/Articles/rio.html.

[20] Robert E. Bieder, *Science Encounters the Indian, 1820–1880: The Early Years of American Ethnology* (Norman: University of Oklahoma Press, 1986).

[21] Johannes Fabian, *Time and the Other: How Anthropology Makes Its Object* (New York: Columbia University Press, 1983), 151.

[22] Muñoz, *Cruising Utopia*, 91.

remade by, a collective modality of endurance and support. It arises from the continuous and difficult work of Indigenous peoples walking toward that opening, or horizon.

Held just before the Rio Earth Summit, the Indigenous conference on the environment at Kari-Oca was attended by representatives from over 200 communities, tribes, and nations from Asia, Africa, Europe, the South Pacific, and the Americas. Together they drafted the Indigenous Peoples Earth Charter addressing Indigenous rights, territory, development strategies, biodiversity, and science. In the Kari-Oca Declaration, they stressed that the authority of Indigenous peoples lay in their positionality, which caused the UN to categorize indigeneity as vulnerable and in need of protection. A powerful statement of collective futurity—"We, the Indigenous Peoples, walk to the future in the steps of our ancestors"—bookended the declaration. And the direct relationship between past and future, evoked in "the footprints of our ancestors . . . permanently etched upon the lands of our peoples," was decidedly spatial, relational, and local. Finally, indigeneity was framed in terms of rootedness: "We cannot be removed from our lands. We, the Indigenous Peoples, are connected by the circle of life to our lands and environments."[23] The declaration claimed authority on the basis of vulnerability and nonuniversal knowledge—a markedly different source than that claimed by the UN.

## The Seventh Generation and Queer Time

In what ways would the principles and temporal orientation in the Kari-Oca Declaration diverge from those in Agenda 21 when enacted? Intergenerational ethics is emerging as a fault line in climate action as the major question shifts from "will we do something?" to "why didn't we do anything?" In 2011, climate ethicist Stephen Gardiner argued that the inability to act is rooted in the climate change problem itself. We have neither an ethical framework nor any precedent to guide us in making choices for future generations on this temporal and global scale.[24] But that fact is simply untrue of most Indigenous epistemologies and ethical frameworks. In his address at the Rio Earth Summit, after the Kari-Oca Declaration was issued, Oren Lyons asserted that if his grandparents had not valued their grandparents' work of transmitting their knowledge, he would not be standing there to commend them or to do the same for future generations. It was his "obligation and responsibility" to ensure that Indigenous knowledge, itself born of generations of relations with nonhumans, would persist, requiring him to link the survival of people, stories, and environment. It is exactly because Indigenous thought is treated as something that must be guarded, and only transmitted through mutually lived experience between generations—both due to its oral form and to histories of colonial termination—that intergenerational ethics are well developed and expressed through time-tested principles in many Indigenous traditions.

For the Haudenosaunee, that principle is known as the seventh generation. After returning home from Brazil, Six Nations scientists and chiefs established the Haudenosaunee Environmental Task Force (HETF). In the same year, the task force published the book *Words That Come before All Else* (1992), on Haudenosaunee TEK. Its authors cite the seventh generation as a source of Indigenous moral authority and motivation for environmental action. Originating in the Great Law of Peace, the seventh generation principle mandates that, even in the smallest of actions, one must think forward to the impact that will be felt seven generations into the future. The source of this perspective is a view of the present as the seventh generation imagined by ancestors, resulting in a long view of history, action, and connection.[25]

In Rio, the UN presented a limited temporal dimension of acceptable environmental risk and could only project a future by positing a clean break from the past, and this schema has continued to dominate discussions of climate change.[26] In his formulation of queer time, Lee Edelman argues that the normative term of US (and arguably Western) political discourse is the value of the child as the ultimate good. Queerness exists in the negative space of that norm, in which there is no claim to the future.[27] This forward-looking, but ultimately limited concern for direct progeny was the normative framework undergirding the declarations made by the UN in 1992. Indigenous temporality and intergenerational ethics subverted that norm at the Kari-Oca Conference, and Haudenosaunee actors have continued this work in their activist and political efforts since then, by proposing an integrated historical view that runs forwards and backwards in time at ever larger scales. That view provides an alternative to the linear and limited model of history that Gardiner and others argue breeds inaction in the Anthropocene. It provides an Indigenous critique, as well, of queer negativity like Edelman's that cannot fathom a positive (and certainly not generative) space outside of settler heteropatriarchy.

---

23   World Conference of Indigenous Peoples on Territory, Environment, and Development, "Kari-Oca Declaration and Indigenous Peoples' Earth Charter," Kari-Oca, Brazil, May 25–30, 1992.

24   Stephen M. Gardiner, *A Perfect Moral Storm: The Ethical Tragedy of Climate Change* (Oxford: Oxford University Press, 2011).

25   Oren Lyons, *A Gathering for the Earth* (National Earth Day Video Conference), Washington, DC: Project Earthlink and Haskell Environmental Research Studies Center (April 21, 1995).

26   United Nations Sustainable Development, "United Nations Conference on Environment & Development, Rio de Janerio, Brazil, 3 to 14 June 1992: Agenda 21," https://sustainabledevelopment.un.org/content/documents/Agenda21.pdf: 154, 226, 241, 267; paragraphs 16, 20, 21, and 22 on biotechnology, toxic chemicals, hazardous wastes, and radioactive wastes.

27   Lee Edelman, *No Future: Queer Theory and the Death Drive* (Durham, NC: Duke University Press, 2004). The severed relationship between futurity and queerness established by Edelman, due to the fact that queers cannot reproduce children, should not be naturalized. Edelman's argument does not imply a situation in which racialized or colonized bodies become normative. Queer and trans people can have children without adopting a straight politic, whereas the fertility and family integrity of Indigenous peoples have been systematically sabotaged by forced sterilization, anti-native family services, and education policies.

## "We don't preach here": Haudenosaunee Epistemic Sovereignty

In 1995, the HETF presented an "Indigenous strategy for environmental restoration" to the UN at the Summit of Elders, which was the first comprehensive Indigenous response to the UN's Agenda 21.[28] In 2002 and again in 2012, they participated in anniversary meetings of the Kari-Oca Conference. They altered the terms under which they interact with settler scientists and academics, forging epistemic partnership models with forestry institutions in Canada and with the Environmental Protection Agency in the US.[29] The HETF and other Haudenosaunee initiatives have also instituted community health studies and education programs that function according to Indigenous research paradigms, and these peograms have served to reinforce efforts for land reclamation, decolonized education, and greater food and health sovereignty.[30] However, HETF scientists, researchers, and spiritual advisors refuse to proselytize.[31]

However briefly, in 1992 the wisdom of the fundamental basis of Western science and the broader claim of modernity—human sovereignty over nature—was questioned. Bill Moyers's depiction of modernity as cityscapes and dominating civilizations, and Boutros-Ghali's view of human history as scientific progress in the conquest of nature, both positioned Western science as the normative structure for all human history and territory. Western epistemic sovereignty functions according to a hierarchy of agents and objects of knowledge. Its internal logic demands theoretical unity and is consolidated through discursive and technological progress. It necessitates competition from within its system; alternative or incompatible concepts are relegated to another sphere (such as religion) or time (TEK).[32] The renewed rhetoric and institutional mode of Haudenosaunee TEK in the 1990s queered Western epistemic sovereignty. In other words, as radical potentiality suspended in colonial pasts and futures—outside the present tense and yet entirely present—TEK subverted the Western scientific demand for competition and stratification.

Native actors at Rio and HETF members in North America had no cause to settle the seeming contradiction between their claims to authority and their vulnerable positionality. Haudenosaunee epistemic sovereignty does not function according to the same ordering of agents and objects. Rather, it situates human agents of knowledge within a lateral network of nonhuman and environmental relations. This was reflected by Oren Lyons's call for representation of the nonhuman at the UN in the 1980s. And it was argued again in the Kari-Oca Declaration, which envisioned a circle connecting Indigenous peoples to nonhuman environments. It also manifested in HETF environmental philosophies that proposed Native-informed environmental restoration projects and Indigenous research methodologies in the late 1990s and early 2000s. Haudenosaunee TEK recognizes natural objects of study without claiming dominion over them. Similarly, its pluralism tolerates alternate claims to sovereignty within the same sphere. Oren Lyons did not fall for Bill Moyers's question of relevancy in 1991 for that reason. TEK's out-of-timeliness does not undermine such knowledge but in fact strengthens its resiliency. Haudenosaunee epistemic sovereignty can be shared and it can be defiantly held, but it can be neither imposed nor proselytized.

\*\*\*

What does a nonhierarchical epistemology and queer temporal position afford when viewing the posthuman landscape of the climate change era? The spatial context for all these discourses is what Ann Stoler and others have called the "ruins of imperial formations."[33] Settler states, former metropoles, and postcolonies are all haunted by the specter of colonialism—the geographic, architectural, epistemic, and historical

---

28  Keith Johnson, Noel J. Brown, and Christopher D. Stephens, "Summit of the Elders: Haudenosaunee Environmental Restoration Strategy," *Akwesasne Notes New Series* 1, nos. 3–4 (Fall 1995): 66–69, https://ratical.org/many_worlds/6Nations/EldersSummit.html.

29  F. Henry Lickers, "Haudenosaunee Environmental Action Plan," *Akwesasne Notes New Series* 1, nos. 3–4 (Fall 1995): 16–17, https://ratical.org/many_worlds/6Nations/HEnvActPlan.html; Mary Arquette, Maxine Cole, Katsi Cook, Brenda LaFrance, Margaret Peters, James Ransom, Elvera Sargent, Vivian Smoke, and Arlene Stairs, "Holistic Risk-Based Environmental Decision Making: A Native Perspective," *Environmental Health Perspectives* 110 (April 2002): 259–64.

30  Elizabeth Hoover, Katsi Cook, Ron Plain, Kathy Sanchez, Vi Waghiyi, Pamela Miller, Renee Dufault, Caitlin Sislin, and David O. Carpenter, "Indigenous Peoples of North America: Environmental Exposures and Reproductive Justice (Commentary)(Report)," *Environmental Health Perspectives* 120, no. 12 (2012): 1645; Carol Jacobs, Cayuga Bear Clan Mother, "Presentation to the United Nations July 18, 1995," *Akwesasne Notes New Series* 1, nos. 3–4 (Fall 1995): 116–17, https://ratical.org/many_worlds/6Nations/PresentToUN.html.

31  Daniel R. Wildcat's *Red Alert! Saving the Planet with Indigenous Knowledge* (Golden, CO: Fulcrum, 2009) provides a counterexample here. But even his promotion of TEK provides a critique of Western modes of political and epistemic authority. Wildcat worked extensively with Vine Deloria Jr. and Steve Pavlik in the early 2000s, and all three advocated Native sciences as an ardent critique of Western scientific hegemony and moral degeneracy. See Vine Deloria, Steve Pavlik, and Daniel R. Wildcat, *Destroying Dogma: Vine Deloria, Jr., and His Influence on American Society* (Golden, CO: Fulcrum, 2006); Vine Deloria, *Red Earth, White Lies: Native Americans and the Myth of Scientific Fact* (Golden, CO: Fulcrum 1995); Vine Deloria, *Power and Place: Indian Education in America* (Golden, CO: Fulcrum, 2001); For more on the politics of Indigenous refusal, see Audra Simpson, *Mohawk Interruptus: Political Life across the Borders of Settler States* (Durham, NC: Duke University Press, 2014).

32  See Michael D. Gordin, *The Pseudoscience Wars: Immanuel Velikovsky and the Birth of the Modern Fringe* (Chicago: University of Chicago Press, 2012). Gordin explores the fact that boundary work—the process of determining what is and is not science (or what is and is not pseudoscience)—is part of the making of scientific knowledge and communities in the first place. It is only in areas where the exclusive authority of science seems under threat that terms like pseudoscience are deployed. Gordin's example covers the popular work of Velikovsky in the 1950s, which had been published by an academic publishing house. Similarly, TEK has been called an "ethnic pseudoscience" (H. David Brumble, "Vine Deloria, Jr., Creationism, and Ethnic Pseudoscience" *American Literary History* 10, no. 2 (July 1998): 335–46). Furthermore, even fairly traditional accounts of how scientific knowledge is derived and transformed, in the history and philosophy of science, assume a competitive framework in which one paradigm must be replaced by another. See Thomas S. Kuhn, *The Structure of Scientific Revolutions* (Chicago: University of Chicago Press, 1970).

33  Ann Laura Stoler, *Imperial Debris: On Ruins and Ruination* (Durham, NC: Duke University Press, 2013).

framework of dehumanization, slavery, and extraction.[34] Ruins and ruinations are spatial and temporal. In this epoch, they exist in the more-than-human and less-than-nature landscapes, and in the discourses on history and time deployed to justify, challenge, and make sense of them. As a potentiality, TEK disrupts the logic that labels the geological horizon as posthuman. Boutros-Ghali's definition of human history at Rio in 1992, which precluded Indigenous action, and the critique of minority rights in relation to nonhuman representation at the UN Working Group on Indigenous Populations both show that not everyone was afforded humanity in the first place. Furthermore, Indigenous epistemic sovereignty exists outside the human-nature dichotomy. Here, the human is not a problem to move beyond. Like the dystopian ruin that has already been a fact of life for Indigenous peoples for centuries, it is a condition that must be inhabited and nurtured so that our voices may carry into the future.

[34] All the various terms proposed to reframe the Anthropocene: the Capitalocene, Necropocene, Plantationocene, Chthulucene, and others engage or resurrect the ghosts of colonialism differently. They highlight the ruinations of capitalism, necropolitical networks, or, in the case of the Chthulucene, the very fact of living with (and within) ruins. They change the way we are meant to understand geology and the meaning of geology in the world around us. See Jason W. Moore, ed., *Anthropocene or Capitalocene? Nature, History, and the Crisis of Capitalism* (Oakland, CA: PM Press, 2016).

"From the Bush to the Table" mural at First Nations School of Toronto, 2017.

If the academy is concerned about not only protecting and maintaining Indigenous intelligence, but revitalizing it on Indigenous terms as a form of restitution for its historic and contemporary role as a colonizing force (of which I see no evidence), then the academy must make a conscious decision to become a decolonizing force in the intellectual lives of Indigenous peoples by joining us in dismantling settler colonialism and actively protecting the source of our knowledge—Indigenous *land*. — Leanne Betasamosake Simpson, 2014

# Wild Life

**Charles Waldheim**

O'Hare Animal Tracks, 1985.

> The environment is not simply "given" . . .
> but is in a crucial sense produced. It is always
> the environment of the system, the outside
> or unmarked space produced by the constitutive
> act of distinction and selection that any system
> uses to secure its operations.
> — Cary Wolfe, *What Is Posthumanism?* (2010)

On the morning of Tuesday, November 16, 2004, a white-tailed deer crossed the airfield at Chicago's O'Hare International Airport and walked into the baggage claim area of Terminal 2.[1] The untimely encounter between human and nonhuman ended badly for the deer yet garnered some attention in the popular press. The story was picked up by the national wire service of the Associated Press, citing a statement from the Chicago Department of Aviation. The *Chicago Tribune*'s transportation reporter published a short piece on the incident titled "Deer Gets into O'Hare but Is Delayed," referring to O'Hare's status as the country's worst airport in terms of on-time performance.[2]

The deer, a young male, was ultimately corralled by airport workers wielding plastic orange construction fencing, and detained until animal control officers arrived. O'Hare employs full-time wildlife biologists from the US Department of Agriculture for just such an eventuality. The wildlife biologists tranquilized the deer in order to remove it for evaluation. They ascertained that the deer had been injured in a collision with a vehicle, most likely outside the airfield. Upon closer examination, the still sedated deer was found to be too badly harmed to be recuperable, and was euthanized.

Press reports described the deer as "bewildered"; as if he were seeking shelter in the airport. The cause of the incident was given as the adolescent buck's bad judgment, which was "clouded by sexual desire" in the midst of the fall mating season. Another explanation: that the deer had been pushed out of his herd by other, "more dominant males."[3]

---

1  Associated Press, "Deer Makes Fatal Detour into Baggage Claim," *Fox News*, November 26, 2004, http://www.foxnews.com/story/2004/11/26/spongebob-kidnapped-plankton-suspected.html.

2  Jon Hilkevitch, "Deer Gets into O'Hare but Is Delayed," *Chicago Tribune*, November 17, 2004.

3  Ibid.

Setting aside the anthropomorphizing tendencies in these accounts of deer sexual practices and politics, the event is worth recounting here as it exemplifies the uncanny encounter between two realms, thought to be discretely described by contemporary systems of urban management.

What is perhaps most illustrative in this untimely encounter between human and nonhuman actors at O'Hare is not the deer's objective that morning (or any speculation thereon), but rather its status in the systems of urban technical management that it traversed. When it was captured and killed, the deer in question was not, strictly speaking, "wildlife." According to the legal and political framework that engineered O'Hare's original construction and ongoing expansion over the past half century, the deer was considered wildlife only until the moment it crossed the security perimeter circumscribing the airfield. Outside that threshold, the deer was understood as an organic, integral element of the "natural resources" or "natural systems" of northeastern Illinois. Inside that threshold, the deer was a biological management problem that had to be minimized in order to secure the safe ongoing operations of a complex system of international aviation.

Not merely an example of the wild (or the natural) entering a space of order and control, at O'Hare wildlife is best described as moving between two systems of modern management. In this traversal, animals leave one realm of highly technical urban management and enter another equally complex technical realm. The subtle semantic distinctions on the boundary between these two managerial systems describe the deer's change of status in legal terms that morning: naturalized wildlife on one side of the line, invasive biological hazard on the other. Press accounts suggested that the disoriented deer followed the interstate highway I-90 onto airport grounds from the neighboring Cook County Forest Preserves. The northern region of the preserve is a wooded area of thousands of acres of regrowth forest habitat along the banks of the Des Plaines River along the eastern edge of O'Hare. The preserve affords ample habitat for deer and other well-adapted urban wildlife immediately adjacent to one of the world's busiest airports.

As early as 1988, several counties in northeastern Illinois began a series of programs to manage or "cull" overpopulated herds of white-tailed deer (*Odocoileus virginianus*). Interventions included the deployment of sharpshooters and the seasonal authorization of hunting to reduce the deer population by the hundreds between November and March of each year; the surplus venison was distributed to local food banks and charities. A spokesperson for the State of Illinois Department of Natural Resources described the deer depopulation program as having two primary managers—hunters and vehicles—as deer-vehicle accidents provided an additional nontrivial amount of deer destruction annually.[4]

The Cook County Forest Preserves and O'Hare's immediate surroundings host other highly adapted urban species deemed significant to airport operations, such as coyotes, raccoons, rodents, raptors, and resident waterfowl. In 1994 alone, for example, two aircraft were damaged by collisions with coyotes on the runway during takeoff.[5] Besides white-tailed deer and coyotes, the biggest wildlife threat to navigable airspace comes from the air. In 2004, the year that O'Hare euthanized our lone bewildered buck, US airports reported over 5,000 potentially catastrophic birdstrikes, or collisions between birds and aircraft.[6]

In 2004, O'Hare retained its status as the busiest airport in the world, measured by total aircraft operations. It was also the worst airport in the US for delays caused by weather, congestion, and the ongoing effects of its outdated runway configuration. In response to these chronic conditions, the city of Chicago published an ambitious masterplan to expand O'Hare's airfield and ease congestion through the construction of a new set of parallel east-west runways to the north and south of the existing airfield. The proposal was titled the O'Hare Modernization Program (OMP). Its implementation required the expansion of the operating airfield by several acres to the north and south. The OMP would enable a more efficient traffic pattern into and out of O'Hare, easing congestion and reducing the number of dangerous live runway crossings by arriving and departing aircraft. Because of its ambitious scale, the OMP required an extensive environmental impact statement as well as authorizations from the Environmental Protection Agency (EPA), Federal Aviation Authority (FAA), and a range of collaborating agencies, authorities, and institutions.[7]

The city of Chicago submitted an environmental impact statement for the OMP in July 2005 and received approval for the project through a Record of Decision issued by the FAA that September. Both the city of Chicago's report and the FAA's assessment refer extensively to the challenges of managing well-adapted urban wildlife species in O'Hare's immediate surroundings. Both documents describe the airfield expansion, the reduction of habitat, and the increased potential for wildlife-airfield interactions. Rather than treat white-tailed deer as "wildlife," however, Chicago's environmental impact statement declared that "deer have been removed from the Aircraft Operating Area"; thus any deer afterwards found within the airport perimeter cannot be considered wildlife. For the purposes of the OMP, deer exist only outside the secure perimeter of the operating airfield. The FAA's approval of the OMP was built upon this distinction and therefore ratified the systemic division between the category of wildlife outside O'Hare's perimeter and biological management inside it.

The ecological implications of the operating airfield at O'Hare illustrate Cary Wolfe's thesis that environments are produced as exteriorities in relation to a particular system.

---

4    Robert Channick, "Deer Population Going Strong, Despite 2 Decades of Culling Programs," *Chicago Tribune*, November 24, 2010.

5    Wildlife Services, *Wildlife Services Assistance at Airports* (Washington, D.C.: US Department of Agriculture, Animal and Plant Health Inspection Service, 2001), 1.

6    Jon Hilkevitch, "O'Hare Enlists Grape Flavoring to Repel Birds," *Chicago Tribune*, November 11, 2004.

7    Federal Aviation Administration, US Department of Transportation, "Appendix N: Biological and Water Resources," *O'Hare Modernization Final Environmental Impact Statement* (July 2005), N-23.

In this instance, the system of airport operations dictates dimensions for the safe takeoff and landing of legions of enormous jet aircraft, crossing immense distances at incredible speeds. The system also specifies precise dimensions of margins for error. These seemingly empty territories are actually technical spaces designed to accommodate the statistically inevitable catastrophe. The modern airfield is engineered to anticipate the aviation disaster and to rapidly remove any incidental human or nonhuman occupation.

Human occupation of the operating airfield is negotiated through strict protocols of security clearance and occupational identification enforced by barbed-wire fences and secured entryways and exits. Nonhuman occupation of the airfield is negotiated through a range of hydrological, ecological, and biological management strategies. Most airports above a certain size employ full-time wildlife biologists as an essential element of their airfield management practices.[8] While much of their effort is directed toward making the airfield territory undesirable to various species prior to any encounter, these biologists and their colleagues are also authorized to capture and kill nonhuman intruders as necessary to maintain the evacuated airspace and its boundary margin of error.

The relative absence of humans on the airfield renders it strangely attractive to a range of species. Yet the planning, engineering, and design of the airfield are conceived to "remove biological function" from its territory. The management and maintenance of the airfield reinforce this approach; major airports engage in a range of practices to present their immediate environments—contained by security fences—as unappealing. This is achieved through the removal of water, food, shelter, as well as through other more aggressive activities. For example, sound cannons and even faux predators are deployed by major airports to enforce the evacuation of biota from the space of aircraft operations. This strange combination of tactics to deter human and nonhuman agents has the counterintuitive effect of making the airfield an even more desirable environment for well-adapted urban species.

Just five days prior to the deer encounter in baggage claim at O'Hare, the *Chicago Tribune* transportation reporter published a lengthy explication on the various ecological and biological strategies deployed by the airport's two full-time wildlife biologists to manage interactions between wildlife and the airfield. To mitigate the constant threat of birdstrikes, for example, they had experimented with the chemical cocktail "methyl anthranilate"—found in grape-flavored Kool-Aid and bubble gum. The national coordinator of airport operations for the US Department of Agriculture describes the chemical compound as a kind of "tear gas for birds." Other ancillary techniques for bird removal included firing sound cannons, draining bodies of water, and the testing of an emergent laser beam technology marketed by BirdTec, Inc., of Hersey, Michigan, as the "Avian Dissuader."[9]

The overwhelming majority of space in the modern airport is given over to "airside" operations, a vast horizontal surface that provides a margin of error for various contingencies, including future expansions; environmental management of adjacent environments; protection for and from surrounding communities and their residents; and the unplanned failure of aircraft operations themselves. This produces an enormous vacuity in the space of the airfield that reads as an undeveloped void. Such spaces must be not only evacuated of human occupation but also, equally significantly, scrubbed of all remnants of previous hydrological and biological processes. Managing the resultant denuded airfield surfaces is a central concern for airport operations, one that also contributes to the sense of estrangement and alienation the airfield engenders in human subjects.[10] Airports tend to be planned for sites as far removed as possible from the cities they serve, though once built, they tend to become the "center" of the metropolitan areas that form around them. This renders the airport's engineered emptiness, and the external environments it produces, central to questions of contemporary urban experience. In this sense, and in its confusion over the status of the ecologies that surround it, the airport is an increasingly significant surrogate for the contemporary city itself.

Aspects of this argument and a reading of airfield ecology appear in the author's forthcoming book, *Chicago O'Hare: A Natural and Cultural History* (Chicago: University of Chicago Press).

8   Sonja Dümpelmann and Charles Waldheim, eds., *Airport Landscape: Urban Ecologies in the Aerial Age* (Cambridge, MA: Harvard Graduate School of Design, 2016).

9   Hilkevitch, "O'Hare Enlists Grape Flavoring to Repel Birds."

10   Charles Waldheim, "Claiming the Airport as Landscape," in *Airport Landscape*, 12–19.

**Image Credits**
143: Courtesy of Robert Burley

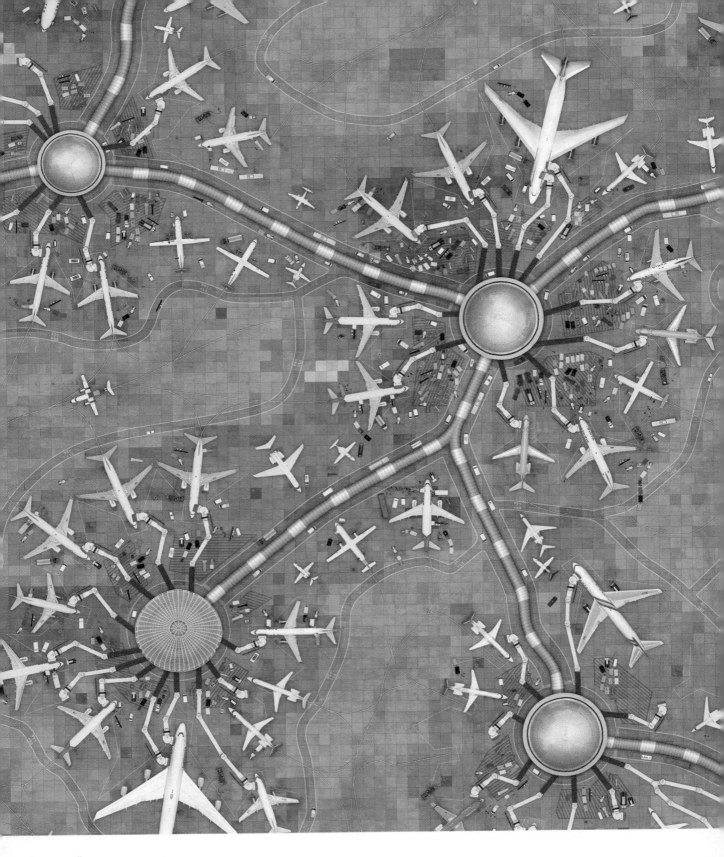
"Aeroporto," by Cassio Vasconcellos.

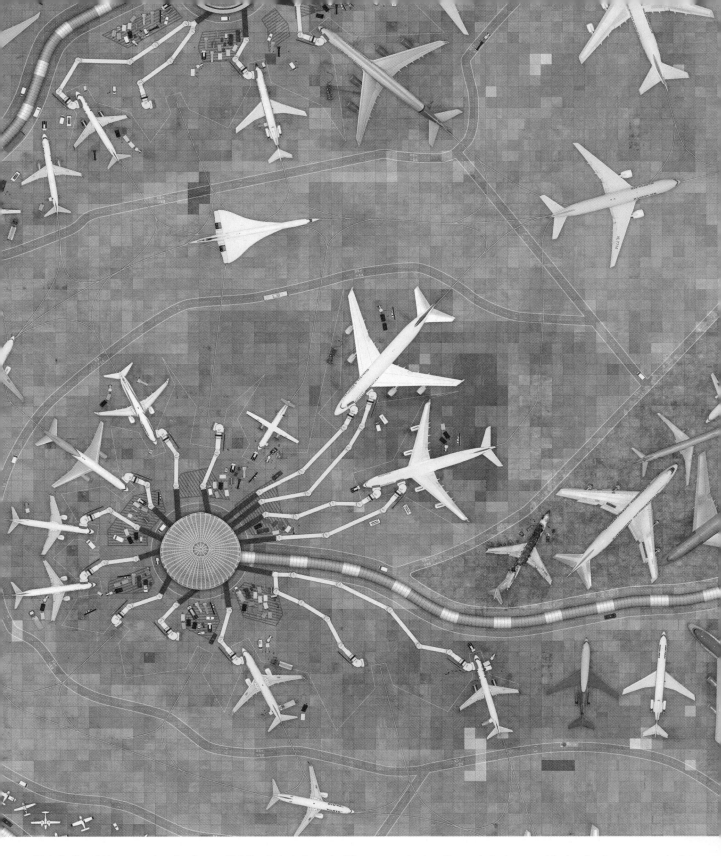

Globalization divides as much as it unites. . . . Mobility climbs to the rank of the uppermost among the coveted values—and the freedom to move, perpetually a scarce and unequally distributed commodity, fast becomes the main stratifying factor of our late-modern or postmodern times. . . . Immobility is not a realistic option in a world of permanent change. And yet the effects of that new condition are radically unequal. Some of us become fully and truly "global"; some are fixed in their "locality"—a predicament neither pleasurable nor endurable in the world in which the "globals" set the tone and compose the rules of the life-game. — Zygmunt Bauman, 1998

# The Cyborg in the Garden

# John Dean Davis

Ahab was a cyborg. At least a low-tech one. His false leg was a constant reminder to the crew of his questionable humanity, symbolic of a corrupting need for vengeance. When Ahab fits his peg into the notch on the *Pequod*'s deck, he joins his organic body with the larger system of the whaleship. Steady in his niche, he commands the gory business of timber and lines, and, as revealed as the journey nears completion, his insanity radiates outward from the quarterdeck, captivating each of the crew in turn. In the first encounter with the white whale, Ahab's peg leg is splintered and he seeks the ship's carpenter for a new one. While the carpenter works, Ahab questions him:

> Carpenter? why that's—but no;—a very tidy, and, I may say, an extremely gentlemanlike sort of business thou art in here, carpenter;—or would'st thou rather work in clay?
> Sir?—Clay? clay, sir? That's mud; we leave clay to ditchers, sir.[1]

The carpenter misses the reference to Genesis, and unwittingly equates the divine creator with the navvy who digs canals. Melville comments here on 19th-century technology and the limits of technical imagination. To a ship's carpenter, there is no great philosophical quandary in making a peg to replace an infinitely more complex biological leg. Familiar with the true lines of a hewn leg—artifice both visually apparent and ontologically untroubled—the carpenter balks at the ability to create with flesh and bone. To manipulate the organic in primordial form and assume the power of a demiurge is unthinkable to the ship's carpenter. He is satisfied with a simplified technological support—the diminished complexity of the artificial leg awakens no desire in this humble representative of the technical class.

I couldn't help but think of the doomed pursuit of the white whale when reading about Wildness Creator, a robot cloud and governing intelligence imagined by Bradley Cantrell, Laura J. Martin, and Erle Ellis, described schematically in a recent article.[2] In their proposal, Wildness Creator, an artificial intelligence (AI) with distributed mechanical agency, is charged with protecting and re-wilding a degraded ecosystem. As its authors imagine, Wildness Creator will evaluate and make decisions about the health of its local dependent wetland, for example, where no humans are allowed to tread. Using drones, it will assist in ecological regulation; it will develop its own algorithmic directives, feeding its intelligence from experience gleaned tending the wetland. Eventually, its authors hope, these actions will become inscrutable to its creators and to other humans, who will be prohibited from interfering at drone-patrolled borders. Wildness Creator then becomes a formidably autonomous guardian. Reading this scenario immediately conjures that other menacing leviathan, bearing scars of previous abuse, patrolling beneath the waves of its patch of the Pacific—its final warning against the *Pequod*'s transgression being to stove her keel.

Wildness Creator is one expression of a revival of the concept of the "cyborg landscape" first introduced to design theory on the heels of Donna Haraway's cyborg manifesto of the early 1990s.[3] However, in the age of ubiquitous Wi-Fi, commercially available drones, do-it-yourself electronics and robotics enabled by Arduino, and intuitive computational engines like Grasshopper, the barriers of feasibility—both technical and financial—have diminished considerably. One no longer has to work for the military or the state to imagine vast sensing networks and remote intervention. Rapid advances and availability of cheap commercial technology have made landscape-scale cyborg a considerably less fantastical creature.

Originally defined as a combination of machine and animal, the cyborg has come to connote a blurrier melding of categories—and can, at various points in the cyborg landscape discourse and literature, mean any combination of biological systems and artificial structures.[4] Increasingly, it is engineered structures and systems of control embedded in "natural" landscapes or ecotopes that are viewed through the conceptual lens of the cyborg. These combinations of "hard" infrastructure and "green" nature are seen critically as examples of asymmetry—indicating the hangover of engineering's historic disregard for the "natural" component of the hybrid.

Robotic assessment, monitoring, and control, particularly regarding city infrastructure, have been the subject of speculation and theorization in urban studies for some decades now.[5] Spreading outward from the city, Wildness Creator is part of an intellectual thrust that identifies opportunity in the power and performance of the urban-as-cyborg, and seeks to imbue non-urban landscapes with sympathetic technology. At its core, the cyborg landscape retains the transgression implied by the merging of organic and machinic—a presumably impure grafting held in suspicion by our culture. Yet there is a sense that this anxiety must be overcome.[6] Landscapes, particularly those replete with degraded and abandoned industrial detritus, are today considered potential sites for saturation by networked, responsive interventions. Technology may have

---

1    Herman Melville, *Moby-Dick, Or, the Whale* (1851; repr., New York: Penguin Classics, 2003), 513.

2    Bradley Cantrell, Laura J. Martin, and Erle C. Ellis, "Designing Autonomy: Opportunities for New Wildness in the Anthropocene," *Trends in Ecology & Evolution* 32, no. 3 (March 2017): 156–66.

3    The first mention of "cyborg" in landscape architecture discourse appears in Elizabeth K. Meyer, "Landscape Architecture as Modern Other and Postmodern Ground," in *The Culture of Landscape Architecture*, eds. Harriet Edquist and Vanessa Bird (Melbourne: Edge, 1994). The recent revival of this direct intellectual lineage is represented by, among others, Bradley Cantrell and Justine Holzman, *Responsive Landscapes:*  *Strategies for Responsive Technologies in Landscape Architecture* (London and New York: Routledge, 2016); and Kees Lokman, "Cyborg Landscapes: Choreographing Resilient Interactions between Infrastructure, Ecology, and Society," *Journal of Landscape Architecture* 12, no. 1 (2017): 60–73.

4    The literature around "natural-cultural" realms is extensive, begun in Bruno Latour, *We Have Never Been Modern*, trans. Catherine Porter (Cambridge, MA: Harvard University Press, 1993). For the purposes of this essay, I invoke "discourse" in a limited sense, meaning recent theoretical discourse in the landscape architecture discipline.

5    See Erik Swyngedouw, "The City as a Hybrid: On Nature, Society, and Cyborg Urbanization," *Capitalism, Nature, Socialism* 7 (June 1996): 65–80; Matthew Gandy, "Cyborg Urbanization: Complexity and Monstrosity in the Contemporary City," *International Journal of Urban and Regional Research* 29, no. 1 (2005): 26–49; Erik Swyngedouw, "Circulations and Metabolisms: (Hybrid) Natures and (Cyborg) Cities," *Science as Culture* 15, no. 2 (2006): 105–21; Matthew Gandy, "The Persistence of Complexity: Re-Reading Donna Haraway's Cyborg Manifesto," *AA Files* 60 (January 2010): 42–44.

6    See Antoine Picon, "Anxious Landscapes: From the Ruin to Rust," *Grey Room* 1, no. 1 (2000): 64–83.

victimized these landscapes in the first place, but we should not shy away from its renewed application in the service of redemption.[7] The machines, having sworn off strip mining and sludge production, stand ready to do the work of restoration. All we need do is find grit enough to endure the irony, and rewrite the governing protocols.

The logic of this narrative, however, remains unsatisfying. Dissonance stems from the gathering obscurity around two prominent intellectual structures in the cyborg landscape's conceptual genealogy. First is the unquestioned dichotomy of biological and mechanical, and the related assumption that any impurity (or even proximity to impurity) produces anxiety in contemporary culture. This problem is closely associated with Leo Marx's image of the machine in the garden—one meant as a point of departure into the deep ambiguity around industrialization and nature in American literature—but increasingly taken as a simple juxtaposition. Donna Haraway's seductive cyborg is the second structure. The brilliant offspring of unholy consummation, the cyborg's transcendent erotics and pleasure in resisting domination give license and absolution to the mingling of sacred and profane. Both concepts are now several decades old and are cited so often and so perfunctorily as to risk loss of definition and effacement of conceptual depth. Revival of the cyborg in the garden ought, at the same time, prompt reassessment of the coupling that produced it.

Leo Marx's *The Machine in the Garden: Technology and the Pastoral Ideal in America*, first published in 1964, was the outgrowth of many years of study of English-language literature focusing on narrative moments of intrusion of technics into pastoral settings. Still widely cited, the work defined an entire scholarly field of analysis. Marx cast a wide net to describe a cultural understanding of the anxieties produced by technological change that swept through the pastoral "middle landscape" in the autumn of the preindustrial world. First defining the Western pastoral in the context of Virgil's poetry, Marx uses *The Tempest* to trace importation of European concepts of the pristine, the wild, and the serenely cultivated to American shores. Marx's work rests on close examination of American literature ranging from Hawthorne, Thoreau, Emerson, and Melville to the cusp of modernism as represented by Henry Adams and William Faulkner. Through subtle analysis of the troubling juxtaposition of industrial technology and nature imagery central to American identity, Marx demonstrates the ambiguities and contradictions of industrialization and its effects on the natural landscape as represented in American art. This defining cultural tension, he argues, produced imagined landscapes of vast and powerful symbolic content.

The key literary motif Marx identifies is the intruding machine—often a locomotive—suddenly interrupting an otherwise peaceful rural repose. The machine causes a sense of "dislocation, conflict and anxiety," by disrupting the self-contained pastoral world characterized by continuity with the past; harmony among human, animal, and cultivated field; and an ancient virtue inherent to pastoral life. Marx marshals a series of similar images drawn from American literature which, in aggregate, constitute "a supreme metaphor of contradiction in our culture."[8] These images reach a sort of culmination at the end of the 19th century, when Henry Adams beheld the dynamos on display at the Paris Exposition and formulated his famous dialectical relation between the "two kingdoms of force" that define an internal civilizational struggle: a preindustrial natural condition, and the visions of infinity revealed by industrial modernity. Marx sees this Calvinist distinction between two "kingdoms" as the product of an ongoing process. The dichotomy is essentially the final stage of mitosis a disentangling of what had previously been entwined—ambiguous or impure layers of machinic and organic that had never before been envisioned as competing worlds, one natural and one not. The 19th century, Marx argues, produced a heterogeneous collection of pastoral notions. Modernist binary thinking tended to emphasize an idealized nature, a pastoral world that Marx identifies as "sentimental and popular." The other world, obscured by modernism, and which he sought to unearth in close readings of works like *Moby-Dick*, was "imaginative and complex"; marked by a comfort with ambiguity that more accurately reflected pre-1900 attitudes.[9]

An "imaginative and complex" pastoralism was in operation when Frederick Law Olmsted built the Back Bay Fens—it is also why landscape historian Elizabeth Meyer pointed to this project when she first articulated the concept of the cyborg landscape.[10] The enduring legacy of Back Bay Fens is Olmsted's achievement of an engineering outcome (solving a sewage problem), while furthering his greater project to change the social space of the city. But to modernist, 20th-century eyes, accustomed to ubiquitous reinforced-concrete control structures, the sinuous lines and prolific vegetation of the Back Bay Fens seemed like anti-engineering. In the context of the mid-19th century, however, the use of plant material and subtle reinvention of topography were the majority of the engineer's tasks on the landscape. Before the widespread use of reinforced concrete in landscape construction in the early 20th century, engineers made do with ancient practices like gabioning, and supplementing the natural angle of repose with hardy rhizomatic grasses—technologies that date to the cultivation of the Nile.[11]

At Back Bay, however, Olmsted broke with the 19th-century metaphysics of engineering. His social mission transcended the mundane but relentless prioritization of control and efficiency that defined engineering practice at the time. A lack of technical ability, resources, or communications

---

[7] For a discussion of the prevalent themes of declension and redemption of nature in Western thought, see Carolyn Merchant, *Reinventing Eden: The Fate of Nature in Western Culture* (New York: Routledge, 2003).

[8] Leo Marx, "Two Kingdoms of Force," *Massachusetts Review* 1, no. 1 (Fall 1959): 77.

[9] Leo Marx, *The Machine in the Garden: Technology and the Pastoral Ideal in America* (Oxford: Oxford University Press, 2000), 5.

[10] Meyer, "Landscape Architecture as Modern Other and Postmodern Ground," 20–21, 25–26.

[11] Efforts to document these ancient practices can be found in technical treatises in Enlightenment France, where vernacular river management and contemporaneous carpentry practice bear fascinating resemblance. See Charles Bossut and Guillaume Viallet, *Recherches sur la construction la plus avantageuse des digues: ouvrage qui a remporté le Prix quadruple proposé par l'Académie Royale des Science, Inscriptions & Belles-Lettres de Toulouse, pour l'année 1762* (Paris: Jombert, 1762); André Guillerme, "La cervelle de la terre: la méchanique des sols et les fondations d'ouvrage de 1750 à 1830," *History and Technology* 7, nos. 3–4 (1991): 211–54.

technology did not prevent 19th-century engineers from imagining (and desiring) expansive regimes of control. In the late 1860s, concurrent with the Back Bay Fens project, American and European engineers installed extensive data-gathering regimes—networks of stream gauge stations—that formed the foundations of continent-wide kingdoms of governance.[12] The state engineering bureaucracies fed on the data streaming from these surveillance networks, forming policies of control and improving efficiency. Though their efficacy of control only reached alarming heights in the mid-20th century, the attitude of the profession toward territory was present from early on, and endures.[13]

It is important to note that 19th-century culture had a markedly different relationship to the natural world than our contemporary culture. Daily life in the 1870s involved many more interactions with animals than we have today, and exposure to harsher elements of weather and disease produced a culture where the precariousness of human civilization in the face of nature was acknowledged more often, and more fully woven into the fabric of life. This would have colored any ontological questions around designing machines that used the forces of nature. Likely, questions of the materiality of the machine would have been subordinated to questions of overall function. Engineers did not agonize over the provenance of the components that formed structures of control. Fashioning interventions that largely mimicked but also sought to realign natural forces, the 19th-century engineer differed little, in terms of the metaphysics of his job, from Melville's carpenter.

The figure of the cyborg, as it emerged in the 1990s, was meant as an avatar of political liberation. The concept was explicitly feminist and socialist—an attempt to engage with technology too ubiquitous and potent to ignore. The provenance of such technology's anatomical components mattered only in that inherent impurity enabled the productive transgression of any confining categories imposed by scientific (i.e., hyper-rational and patriarchal) scrutiny. Citing the manifesto merely in support of questions of material ontology—hybrid mixing of the stuff that constitutes landscape, both biotic and abiotic—neglects its original, radical mythology. Haraway's conception of the cyborg as "illegitimate offspring" of militarized science and a gendered, painfully othered nature finds the basis for liberation in its illegitimacy. Unconcerned with adherence to the categories of an aggressive and exploitative patriarchal science, the cyborg is free to take pleasure in its inscrutability; its inherently distorted nature is also a shield against regimes of totalized control.

Cyborgs are also fundamentally immaterial: "cyborgs are ether, quintessence" and their media is language.[14] Bearing the influence of the academic discourse of semiotics from the late 1980s, Haraway's concept is nonetheless relevant to our contemporary situation, where talk of "coding" and the scripting of cultural routines is entirely ubiquitous. The cyborg can speak to machines; it is fluent in the code of programming languages and can anticipate the need for procedures. It finds uncanny comfort in a machine epistemology driven toward more definition, more individuation, and more control. However, the cyborg's animal nature means that although it speaks the idiom of machines, as well as a million unrecognizable others. This bestows "a powerful infidel heteroglossia" on the cyborg that spirals out of the machine's realm of thought and creates new territories of meaning beyond the hardened categories of militarized science. Incomprehensibility becomes the cyborg's refuge and instrument of resistance, allowing it to disrupt those communications on which the dominating structures of control rely. "Cyborg politics," Haraway notes, "insist on noise and advocate pollution, rejoicing in the illegitimate fusions of animal and machine."[15]

In the context of the Cold War, Haraway's suspicion of ubiquitous militarized command-and-control structures was not abstract. Continental networks of distributed weapons systems of awesome destructive power remain the greatest threat to life. Invoking La Malinche, the ambiguous figure who was the lover of Cortés and the first translator between Nahuatl and Spanish, Haraway sketches the cyborg's shadowy territory of survival in the liminal zone between life and death. Perceiving that the Spanish and their language represented coming death and destruction, La Malinche spoke the needful words to ensure her children's survival. In her case, comfort with impurity did not negate or remove the very real dangers connoted by the language of Cortés and his men.

It is difficult to see the sensor-laden and drone-tended landscape as fulfilling the truly radical promise of the cyborg. It is much easier to see a militarized command-and-control apparatus—the "machine as gardener"—as another manifestation of the sins of 20th-century engineering: a monumental effort at totalized control. Indeed, the Wildness Creator purports to prioritize biological needs over human needs. But it remains a monolithic intelligence with regional agency. Typologically, it is more similar to the Hoover Dam than to an actual wetland, which has no governing intelligence. Though distributed over a vast area and semi-hidden in the mud, it remains fundamentally monumental.

It strikes me as unfortunate that contemporary design theory seeks to instrumentalize powerful concepts, including the cyborg (though not exclusively Haraway's notion), in an effort to make landscape architecture more akin to engineering. The current appeal to create cyborg landscapes is

---

12  See United States, *Annual Report of the Chief of Engineers to the Secretary of War for the Year 1870* (Washington, DC: Government Printing Office, 1870), 401–04; Giacomo Parrinello, "Charting the Flow: Water Science and State Hydrography in the Po Watershed, 1872–1917," *Environment and History* 23, no. 1 (February 2017): 65–96.

13  Emergence of the idea of control was by no means static, and conceptions of how to make environments "responsive" have undergone a fascinating course of evolution. See Antoine Picon, "L'avènement de la ville intelligente," *Sociétés* 132, no. 2 (2016): 9–24.

14  Donna Haraway, "A Cyborg Manifesto: Science, Technology, and Socialist-Feminism in the Late 20th Century," in *The Cybercultures Reader*, eds. David Bell and Barbara M. Kennedy (London: Routledge, 2001), 312.

15  Haraway's argument revolves around the notion that "informatics is domination," and that "the translation of the world into a problem of coding" represents the fundamental hegemony of technical language, and the ideal network—that of internal communication and control—to interrogate critically. See ibid., 300–04.

fundamentally to prioritize certain landscapes that, in their form and original intent, and in their arrays of processes that transform materials, is largely indistinguishable from the engineer's priorities. To reduce landscape architecture to engineering would be to make the same affirmative mistake as Melville's ship carpenter. A myopic stubbornness leads to ossified ways of working that also have a historically consistent metaphysics. Though it would require a profound shift, why not reject the professional legacy of skilled "carpentry" and instead imagine what it could be like to work in clay?

The schematics of Wildness Creator contain a gesture toward answering Ahab's invitation. I remain skeptical of its ability to learn its way out of being, fundamentally, a monolithic intelligence, and doubt any human-designed machine will ever be able to truly exist as an ecosystem. But the notion that Wildness Creator could learn to assemble "emergent" patterns and processes does indicate a pathway into a post-human environmental ethos—it poses Ahab's invitation as a matter of thought.

Engineering and the design professions have specifically defined means of measuring and interpreting the world, which necessarily shape the cascade of decisions made and narrow each field's scope of allowable thought. This is the carpenter's problem; in maturity, the paradigm of engineering the natural world is to seek more and more finely grained control, a desire that has been relatively continuous in practice since the profession began.[16] It is tightly bound to both success in applying increasingly powerful scientific knowledge, and the humanist narratives that valorize mastery and spin soothing stories of cultural progress. The carpenter's permissible action and boundaries of thought need certain comparative structures to mark advancement—perhaps this is why the nature-culture dichotomy remains so stubbornly entrenched. Satisfying this narrative, however, produces a scenario in which industrial capitalism pursues a harmonious relationship with wetlands, a possibility that is both increasingly desired and increasingly untenable. The carpenter is too constrained by the tools and instruments with which he gains measure of the problem.

Perhaps, like Ahab, we need to discard our navigational instruments before we venture further. Post-engineering thought might be imagined useful somewhere other than the governing protocols of yet another machine. Perhaps, as a model, we should heed Eduardo Kohn's memorable prompt to "think like a forest." This may mean eschewing language or other imposed notions of "coding" and first listening to the natural world, differently. "To engage with the forest on its terms," Kohn writes, "to enter its relational logic, to think with its thoughts, one must become attuned" to its forms, processes, and patterns of propagation—these pathways promise the shape of a nascent epistemology.[17] Dreaming of a post-engineering world is both an invitation to create new instruments and then new notions, and not to preclude that "clay" technology may be radically different from what is already known. So different, indeed, that the 20th-century engineer's interventions in the natural world will seem crude by comparison.

The approach of a truly posthuman machine, one thought by a forest or wetland, inclusive of all its inhabitants (humans and non), would present no dichotomy, nor would it be seen as an intrusion at all.

---

16  See Thomas S. Kuhn, *The Structure of Scientific Revolutions* (Chicago: University of Chicago Press, 1970), 110–11.

17  Eduardo Kohn, *How Forests Think: Toward an Anthropology Beyond the Human* (Berkeley: University of California Press, 2013), 20.

Coal terminal, Port of Qinhuangdao, China, 2015.

[A] cyborg world is about the final imposition of a grid of control on the planet. . . . From another perspective, a cyborg world might be about lived social and bodily realities in which people are not afraid of their joint kinship with animals and machines, not afraid of permanently partial identities and contradictory standpoints. The political struggle is the will to see from both perspectives at once. — Donna Haraway, 1984

# Horizon of a Different Machine: Geotechnicity

## Namik Mackic & Pedro Aparicio Llorente

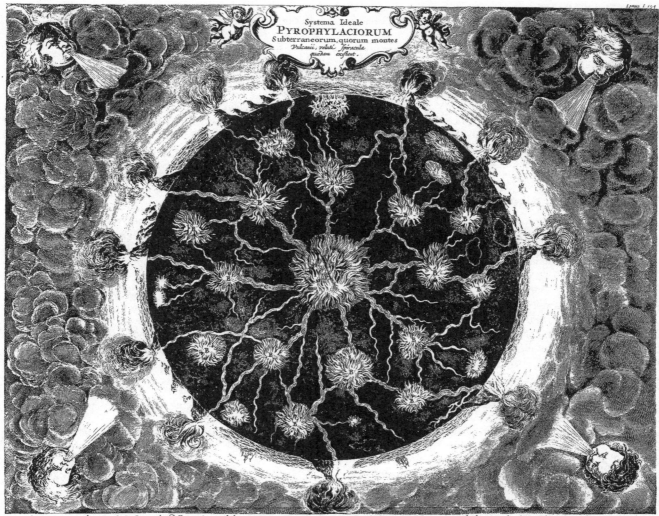

System of subterranean fires, from Athanasius Kircher, *Mundus Subterraneus* (1678) vol. 1, 194.

*I remember when I was small. When the sea met some jellyfish. The jellyfish were translucent, with no heart, spleen, liver, nor intestines. There was only water, water, and water filling their form. I thought, how beautiful it would be if my physique could be like that. How beautiful if I could put all the fish in the sea into my self. Also the coral and the algae—at night I would shine a flashlight on my body.*
— Mardi Luhung, *Teras Mardi* (2015)

## The Conveyor

Conduct an internet search for "subduction" and scroll through the image results: an array of brightly colored section diagrams offers the same schematic view onto "the dominant physical and chemical system of Earth's interior . . . our planet's largest recycling system."[1] When one tectonic plate slides under another, the oceanic crust grinds into a conveyor belt of matter that plunges toward the viscous asthenosphere for millions of years until it is discharged to the surface by orogenic processes and pyroclastic flows. We live inside this perpetual machine; mantle convection supports life, tectonically: "One can speculate that if subduction zones did not exist to produce continental crust, the large exposed surfaces of rock known as

---

1     Robert J. Stern, "Subduction Zones," *Reviews of Geophysics* 40 (2002): 3.

continents would not exist, the earth's solid surface would be flooded, and terrestrial life, including humans, would not have evolved."[2]

This tellurian metabolism is revealed through the topographic event that we call "volcano," but the intricate ways in which its formations ripple through our world are not readily grasped. If the specificity of landscape lies in the potential for particular technologies to emerge, then "the volcanic" marks an entry point into an evolutionary story of technics in which the human has been forged and propelled forward, not once occupying its heroic narrative center.

## Two Geneses

> Architecture could only find its place after the Flood—or rather, in its stead.
> — Hubert Damisch, *Noah's Ark: Essays on Architecture* (2016)

In the digital collections of the Linda Hall Library, "the world's foremost independent research library devoted to science, engineering, and technology," a first-edition copy of *Arca Noe* (1675), by 17th-century German Jesuit polymath Athanasius Kircher, attracts notice.[3] The book's sequence of engravings lays out the construction of Noah's Ark. Species of exotic birds, reptiles, and both feral and domesticated mammals—many gleaned from descriptions dispatched from the New World by Kircher's brethren[4]—are sequestered into a rectilinear grid plan, and rescaled to fit the cell's dimensions, which only accentuates the abstraction of the scheme. The boarded ark is rendered in a three-story plan oblique, sliced open to display each story as divided by a central corridor with symmetrically arranged units on either side. The architecture references ship typologies of the Age of Discovery: caravels, carracks, *fluyts*, *retourschepen*—those floating extensions of colonial town-making that proto-urbanized the ocean. The plan exposes the episteme of the colonial project, structuring what Foucault names "the taxonomic area of visibility."[5] Kircher's ark carries a typological arrangement that strikingly projects into the future, linking the colonial fixation on immunized space to one of its modern conclusions—social engineering through standardized dwelling: *une maison est une machine-à-habiter*.[6]

Depictions of Noah generally advance the hierarchy of man over machine. An illuminated, probably 15th-century manuscript by an unknown author shows Noah rising through the oculus in the ceiling of a spherical capsule, his head and hands outlined against the celestial vault, while livestock crowds the belly of the ship. Here is man wielding biopower *avant la lettre*: he has built the boat, he has cared for the animals. Now the dove he has released returns with a freshly plucked olive branch in its beak, announcing a restored earth—a *terra nullius*—ready to be colonized. If Noah's Ark epitomizes the contemporary, technomanagerial paradigm of the urban, marked by an expanded mandate to contain contingency, now reconceptualized as "risk," it is because it carries the embryo of techno-utopia, the positivist dream of replacing a volatile material reality with a technologically stabilized one. How can we imagine habitation outside and beyond the predictable failure of this scheme, when all we see is another extinction event, another deluge rising?

A decade before *Arca Noe* was published, Kircher revealed a vision of the world in *Mundus Subterraneus* (1665) that radically departs from this anthropocentric scheme. In one of the earliest attempts to divulge the earth's interior by means of a cutaway diagram, a circuitry of flows burrows through the planetary body, depositing "metallic and mineral juices" in its wake. Conceived by Kircher as a "scattered collection of historical relations,"[7] this underworld is described as

> a well-framed house, with distinct rooms, cellars, and store-houses, by great art and wisdom fitted together; and not as many think, a confused, jumbled heap or chaos of things, as it were, of stones, bricks, wood, and other materials, as the rubbish of a decayed house, or a house not yet made.[8]

Kircher's written treatise is bound up in theological telos and allusions to absolute unity, yet his visuals depict emergence, metamorphism, unstable systems.[9] This planetary architecture breaks with the dominant schema of Aristotelian

---

2   Ibid., 3.

3   Athanasius Kircher, *Arca Noe, in Tres Libros digesta, Quorum I. De rebus quae ante Diluvium, II*.

4   As professor for more than 30 years at the Roman College—the Jesuit base in the Vatican used to educate priests who would be sent across the colonial globe—Kircher was exposed to an ever-expanding inventory of plants and animals registered by Jesuit missions. This diversity of species deconstructed biblical paradigms, leading Kircher to intuit certain principles of evolution.

5   Michel Foucault, *The Order of Things* (New York: Pantheon, 1970), 137.

6   "The land-dweller's house is the expression of an outdated world of small dimensions. The [ocean] liner is the first stage in the realization of a world organized in accordance to the new spirit." Le Corbusier, *Toward an Architecture* (1923; repr., Los Angeles: Getty Research Institute 2007), 158. Jean-Louis Cohen contextualizes Le Corbusier's statement: "[His] fascination with ocean liners, the first class of objects taken up in his ocular triad, long predated his writing of *Vers une architecture*. . . . The ocean liner is what inspired Le Corbusier to create the provocative aphorism: *La maison est une machine à demeurer* ('The house is a machine for residing'). In the 1924 version, *demeurer*, with its static and bourgeois connotations, is replaced by *habiter*, 'to live in' or 'to inhabit.'" See Jean-Louis Cohen, "Introduction" in Le Corbusier, *Toward an Architecture*, 14.

7   Athanasius Kircher, *Mundus Subterraneus, in XII Libros digestus; quo Divinum Subterrestris Mundi Opificium, mira Ergasteriorum Naturæ in eo distributio, verbo pantámorphou Protei Regnum, Universæ denique Naturæ majestas & divitiæ summa rerum varietate exponuntur. Abditorum effectuum Causæ acri indagine inquisitæ demonstrantur; cognitæ per Artis & Naturæ conjugium ad Humanæ vitæ necessarium usum vario Experimentorum apparatu, necnon novo modo & ratione applicantur* (Amsterdam: Apud J. Janssonium & E. Weyerstraten, 1665). English edition: *The Vulcano's, or, Burning and Fire-vomiting Mountains, Famous in the World: With their Remarkables. Collected for the most part out of Kircher's Subterraneous World; And exposed to more general view in English, upon the Relation of the late Wonderful and Prodigious Eruption of Ætna. Thereby to occasion greater admirations of the Wonders of Nature (and of the God of Nature) in the mighty Element of Fire* (London: J. Darby, for John Allen, 1669).

8   Kircher, "The Explication of the Schemes," in *The Vulcano's*, I.

9   Kircher's writing "calls upon religious and hermetic philosophies in his interpretations of the subterranean word." See William C. Parcell, "Signs and Symbols in Kircher's *Mundus Subterraneus*," in *The Revolution in Geology from the Renaissance to the Enlightenment*, ed. Gary D. Rosenberg (Boulder, CO: Geological Society of America, 2009), 63–74.

hylomorphism, arising instead from the morphogenetic capacities of matter. What the illustrations most consistently foreground is fire, which commands the distribution of "life and heat" to all the earth. Fire is integral to both *oikos* and *nomos* in what Kircher refers to as an "internal Oeconomy."

Kircher advances fire as something that approximates, in Deleuzian terms, a probe-head of the earth's machinic phylum: a rhizomatic agent that ruptures strata and sets morphogenetic processes in motion.[10] While *Arca* delivers a prescient parable of what currently transpires before our eyes—the full-scale reterritorialization of life by technology—in *Mundus*, Kircher anticipates the late modernity's recognition of the nonlocal character and effects of environmental transformation[11] as something already detectable in the dynamic stratigraphy of large-scale geomorphic processes:

> You see how Water, Fire; Fire, Water; mutually, as it were, cherish one another; and by a certain unanimous consent, conspire to the Conservation of the Geocosm, or Terrestrial World. For if Subterraneous Fire should emit no vapors for matters of Winds; The Sea, as it were torpid, and void of motion, would go into a putridness, to the ruin of the whole Globe.[12]

We now begin to appreciate the ambiguous status of the human in Kircher's later work. His container-shaped ark is without bow or stern—the trajectory of its passage is for the waves to steer. On land before the flood—and during it, inside the hull—a multitude of human figures labors in the service of a regime that substitutes a schematized inventory for geo-ecological continuum: a zoo for *zoe*. A before-and-after pairing of section views shows how the ark is lowered by the receding water and caught by a protruding landform, much like a prefab unit landing on a construction site.[13]

Two distinct and complementary ontologies that regard technics as native to seemingly inanimate matter extend from Kircher's work. The first ontology foreshadows technology as it was theorized in the 20th century; a self-propagating system of functional substitutions for natural phenomena, which picked up momentum during the long period of European colonialism. Within this concrescence, the human subject is at once propelled and held captive by technical schemas that combine and multiply, which subsume its lifeworld into the fragmented horizon of their endless iterations—a spell of virtual futurity. The other ontology points to the preexisting, interlaced substrate of geophysical agents that harbors an immanent technicity of its own; as it moves materials around—including the human appropriation of, and intervention in, these movements' patterns—the morphogenetic potentialities multiply.

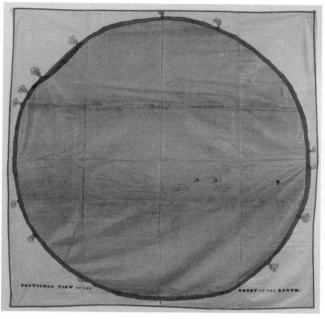

Sectional view of the crust of the earth, by Orra White Hitchcock (1796–1863).

## Topography of Technics

In *Du mode d'existence des objets techniques* (*On the Mode of Existence of Technical Objects* [1958]), Gilbert Simondon establishes technicity as "one of the two fundamental phases of the mode of existence of the whole constituted by man and the world," the other phase being religious thought.[14] The preeminent 20th-century Western philosopher of technics names these as two cardinal modes through which man mediates his material existence in the world, and he locates their separation historically, in a "phase shift," or rupture in a "primitive magical unity" between the human and the universe:

---

10  Deleuze and Guattari introduce the probe-head in their elaboration of the 'faciality machine'—a concept that encapsulates the territorialization of the human body by the head. See Gilles Deleuze and Félix Guattari, *A Thousand Plateaus: Capitalism and Schizophrenia*, trans. Brian Massumi (Minneapolis: University of Minnesota Press, 1987), 190.

11  Theorist Timothy Morton proposes "non-locality" as a defining characteristic of "hyperobjects"—objects "massively distributed in time and space such that any particular (local) manifestation never reveals the totality of the hyperobject." See Morton, "Hyperobjects Are Nonlocal," *Ecology without Nature*, November 9, 2010, http://ecologywithoutnature.blogspot.com/2010/11/hyperobjects-are-nonlocal.html, and Timothy Morton, *Hyperobjects* (Minneapolis: University of Minnesota Press, 2013).

12  Kircher, "The Explication of the Schemes, out of Kircher," in *The Vulcano's*, I.

13  "For 40 days the flood kept coming on the earth, and as the waters increased they lifted the ark high above the earth. The waters rose and increased greatly on the earth, and the ark floated on the surface of the water." *Zondervan NIV Study Bible*, ed. Kenneth L. Barker, full rev. ed., (Zondervan, 2002), Gen. 7.17–18. "At the end of the 150 days the water had gone down, and on the 17th day of the 7th month the ark came to rest on the mountains of Ararat." (Gen. 8.3–4)

14  "By phase, we mean not a temporal moment replaced by another, but *an aspect that results from a splitting in two of being and in opposition to another aspect*; this sense of the word phase is inspired by the notion of a phase ratio in physics; one cannot conceive of a phase except in relation to another or to several other phases; in a system of phases there is a relation of equilibrium and of reciprocal tensions; it is the actual system of all phases taken together that is the complete reality, not each phase in itself." Gilbert Simondon, "The Genesis of Technicity" excerpted from *On the Mode of Existence of Technical Objects*, trans. Cécile Malaspina and John Rogove (Minneapolis: University of Minnesota Press, 2016). Also published in *E-Flux Journal* 82 (May 2017), https://www.e-flux.com/journal/82/133160/the-genesis-of-technicity/. All subsequent Simondon quotations are adapted from the online source.

We suppose that the primitive mode of existence of man in the world corresponds to a primitive union, prior to any split, of subjectivity and objectivity. The first structuration, corresponding to the appearance of a figure and a ground in this mode of existence, is the one that gives rise to the magical universe. The magical universe is structured according to the most primitive and meaningful of organizations: that of the reticulation of the world into privileged places and privileged moments.[15]

Simondon speculates that this topology, in which certain spatially and temporally defined features are privileged over others, has prepared the ground, so to speak, for the emergence of technics. It is "as if all of man's power to act and all the world's ability to influence man were concentrated in these places and in these moments." Remapping the prehistoric physical world onto a spiky cognitive geography of the early humans, Simondon effectively proposes that certain particular, prominent, or idiosyncratic topographical features have comprised "key-points" of evolutionary importance to the early human, which in turn have translated into the "schematisms" of the very first technical objects. In speculating on the significance of this reticulated universe for the early human, Simondon's theory reads almost like the ethnography of some volcano-centric culture.

> A privileged place, a place that has a power, is one that drains from within itself all the force and efficacy of the domain it delimits; it summarizes and contains the force of a compact mass of reality; it summarizes and governs it, as a highland governs and dominates a lowland; the elevated peak is the lord of the mountain . . .

And here a footnote interjects: "Not metaphorically, but really: it is toward it that the geological folding orients itself and the push that has edified the entire high plateau."

Simondon has been hailed as a lucid and generous examiner of the prospect of "humanity contained in the machine."[16] In setting up "the whole constituted by man and the world" as the proper dispositif for his inquiry into technics, however, he presupposes that technics originated along the surface of a planet already populated by humans. He asks: "What could be the motor of the successive splits [between the human and the world] in the course of which technicity appears?" To which we reply: What if there is a domain or manifestation of technicity that is integral to the coming-in-to-being and mode of existence of "the whole," that is prior to, and independent of, the emergence of the human?

Our proposal that earth systems are imbued with modes of technicity of their own invokes seemingly unorthodox yet key criteria of conventional technology. Simondon paves the way for recognition of the particular vitality of technical objects: in his terminology, they undergo "genesis" from within a "pre-individual field," toward becoming "individuated" objects. The very incompleteness of technical objects—their heteronomy—is what fecundates technical schemas: it grants them the capacity to spawn integrative systems via recursive cycles of prototyping and optimization. Incompleteness distinguishes technical objects from living beings, but accords them nonetheless certain characteristics of life-forms. Naming this process *concréscence* ("concretization"), Simondon conjures a *technogenesis* in which humans are *operators* of technological evolution, rather than its proper *originators*.[17]

If the planet as a whole represents an incomplete, ruptured, and liquefied ground that generates technical schemas available to be picked up and realized by humans and other species, then this is precisely what renders it vital—not in the sense of some sentient, self-regulating "whole," but as the bearer of a latent technicity of its own.[18]

When Simondon appropriates the concept of "mediation" to account for the relay, or take-up, of topographic schematisms in primitive technical objects, he locates a crucial link between geogenic and anthropogenic formations of matter. These mediations may have a tendency to materialize into technical objects that detach from their ground of origin—but they do not exhaust the forces and pressures that trigger mediation in the first place: these forces persist, as vectors animating the landscape. If regional geomorphic—more specifically, volcanic—phenomena have provided the decisive adaptive threshold that set early hominins on their evolutionary journey,[19] then the geophysical substrate as a whole, and in its myriad regional particularities, constitutes one continuous mesh of capacities to spawn alternative technogeneses.

---

15   "While, in the magical reticulation of the world, figure and ground are reciprocal realities, technics and religion appear when figure and ground detach themselves from one another, thereby becoming mobile, fragmentable, displaceable, and directly manipulable because they are not bound to the world. *Technical thought retains only the schematism of structures, of that which makes up the efficacy of action on the singular points; these singular points, detached from the world whose figure they were, also detached from one another, losing their immobilizing reticular concatenation, become capable of being fragmented and available, as well as reproducible and constructible.* The elevated place becomes an observation post, a watchtower built on the plain, or a tower placed at the entrance of a gorge." Simondon, "The Genesis of Technicity;" our italicization.

16   John Hart, "Preface" in Gilbert Simondon, *On the Mode of Existence of Technical Objects*, trans. Ninian Mellamphy (London, ON: University of Western Ontario, 1980), i.

17   Philosopher Bernard Stiegler summarizes this aspect of Simondon's thought thus: "In the industrial age, the human is not the intentional origin of separate technical individuals qua machines. It rather executes a quasi-intentionality of which the technical object is itself the carrier." Bernard Stiegler, *Technics and Time, 1: The Fall of Epimetheus*, trans. Richard Beardsworth and George Collins (Stanford: Stanford University Press, 1998), 67.

18   Speculating on the mechanics of what later would be known as mantle convection, Kircher acknowledges that events manifesting in one spot on the surface of the earth have their origin in a movement triggered deep within it. Kircher's interior rivers of fire divulge a world dissipating at its core, not a hyperobject so much as a hyperevent.

19   Geographer Michael Medler argues that some 1.8 million years ago, a massive, 200,000-year-long lava flow in the Olduvai Gorge provided consistent sources of fire—and "very specific adaptive pressures and opportunities"—to small isolated groups of hominins. The occurrence of the flow coincides with the period in which *Homo erectus* emerged, distinguished from earlier fossil humans by bipedalism, shorter intestines, larger brains, and smaller teeth and mouths—all of which are fire-specific adaptations. Medler's theory suggests that hominins began adapting to fire some 800,000 years before the earliest evidence of its controlled use. See Michael J. Medler, "Speculations about the Effects of Fire and Lava Flows on Human Evolution," *Fire Ecology* 7, no. 1 (2011): 13–23.

Simondon's concept of reticulation now becomes useful as a conceptual precedent for detailing regional expressions of geotechnicity (we might provisionally describe these as the *longue durée* of geomorphic scripts native to a region). Thus, when recounting that magical, supposedly pre-technical universe, Simondon unwittingly tells of webbed and generative geo-human relations that today can still be observed in certain landscapes marked by relatively high geological, seismic, volcanic, or hydrogeomorphological activity:

> The magical world is made of a network of places and of things that have a power and that are bound to other things and other places that also have a power. . . . These places and these moments keep hold of, concentrate, and express the forces contained in the ground of reality that supports them. [They] are not separate realities; they draw their force from the ground they dominate; but they localize and focalize the attitude of the living vis-à-vis its milieu. . . . Each singular [key-]point concentrates within itself the capacity to command a part of the world that it specifically represents and whose reality it translates, in communication with man.[20]

We suggest that these "points," which in a situated material practice evolve into coherent fields of mediation, are privileged in the imaginary of the early human precisely *because of* the adaptive pressures and opportunities that they contain in the sum of their physical properties: they attract, constrain, enable, and guide "man's insertion into natural coming-into-being." Further on, Simondon concedes: "One could call these singular points the key-points commanding over the man-world relation, in a reversible way, for the world influences man just as man influences the world."

We now begin to recognize how the geomorphic performance of the earth systems, expressed across a continuum of processes that links the planet's interior with its surface and envelope, historically gives rise to a progressively self-divergent evolutionary column of technics, as mediated through institutionalized or socialized practices of the human species. One "phase" of this column (to borrow Simondon's definition) has produced the *bounded technicity* of industrial systems, beginning with the military adaptation of various artisanal techniques and evolving into the insular platforms of large technological systems that have enlisted science in the propulsion of their iterative phases. The other phase of the column may still be observed in those endogenous paradigms where regional vectors of geotechnicity are actualized, and even socialized, in the material practices of everyday life. What if there were dynamic cartographies capable of capturing the links and relationships between the substrate and the surface of this latter paradigm, that is, the vectorial systems of landscape dynamics and material practice within which everyday life is lived?

## Mapping the Volcanic

> In front of me I had put on display symbols of my faith: a terrestrial and celestial globe, a sextant, artificial horizon, telescope, chronometer, thermometer, psychrometer, a compass, magnet, microscope, Nicholson's airometer, a triangular prism, portable camera obscura, a daguerreotype camera, a small chest for doing chemical tests, and other such tools of applied science.
> — Franz Wilhelm Junghuhn, *Light and Shadow from Java's Interior* (1854) [21]

Seen from above, the temple Pura Luhur Poten is a fleck in a sea of sand, a split gate followed by three consecutive courtyards one kilometer away from the topographic and cosmological center point of the Tengger identity: the crater of Mount Bromo in East Java. From the temple, dirt roads unravel along ridges and through villages nested in between intensively cultivated terraced slopes that spill into the lowlands. On day 15 of the last month of the Tengger year, the Kasada pilgrimage gathers villagers around the volcano's rim. Blessed produce, livestock, and money, arranged in pyramidal offerings, are carried up and thrown down into the abyss, honoring the mountain that supports them, the "refugees of the Majapahit." [22]

Linguistic anthropologist Nancy J. Smith-Hefner notes the first mention of Kasada in a 500-year-old bronze plate from the Majapahit court, alluding to the Tengger as "the spirits' servants of the honored holy mountain Bromo." [23] The inhabitation of an active geology rests upon such relational synchronization of human and non-human identity. Anthropologist Robert Hefner refers to this regional ritual as a *field of activity*[24] that replaces an apparatus of ossified institutions with

---

20   Gilbert Simondon, "The Genesis of Technicity."

21   This publication reveals Junghuhn's opus—in his lifetime recognized as an important scientific contribution—as a project of theo-technological determinism. For a review of Junghuhn's literary figures in the context of colonialism, see Robert Nieuwenhuys and E. M. Beekman, *Mirror of the Indies: A History of Dutch Colonial Literature* (Hong Kong: Periplus, 1999), 67.

22   The Majapahit Empire was a thalassocracy based in Java between the late 13th and early 15th centuries. Its vast archipelagic expanse is the precursor for modern Indonesia's boundaries. The fall of the Majapahit gave rise to the first Islamic sultanate in Java. The remaining Majapahit retreated to the mountains of East Java and Bali. In his work on Zomia, political anthropologist James C. Scott refers to the Tengger "hilly refugees from state control" and to Kasada as commemoration of autonomy: "The history of flight is remembered annually by the non-Islamic highlanders who throw offerings into the volcano in remembrance of their escape from Muslim armies. Their distinct tradition, despite its Hindu content, is culturally encoded in a strong tradition of household autonomy, self-reliance, and an anti-hierarchical impulse." See James C. Scott, *The Art of Not Being Governed: An Anarchist History of Upland Southeast Asia* (New Haven, CT: Yale University Press, 2011), 134–135.

23   Nancy J. Smith-Hefner, "Language and Social Identity: Speaking Javanese in Tengger" (PhD diss., University of Michigan, Ann Arbor, 1983), 11.

24   Robert W. Hefner, *Hindu Javanese: Tengger Tradition and Islam* (Princeton: Princeton University Press, 1990), 44–46.

no regional division of labor that might link villages in the exchange of specialized goods and services, . . . no clans, castes, or other king groups to provide intra-regional corporate organization . . . ritual provides a motive and an organization for social interaction among people from diverse Tengger communities [and] the maintenance of a sense of shared identity.[25]

That identity is not only performed during Kasada, it is practiced in the way the topography itself is appropriated as a compass that triangulates mountain, urban form, and human body:

> The villager situates himself according to the mountain incline; *isor* "under, below or at the bottom" or *dhuwur* "above, at the top." . . . The directions in each Tengger village are thus relative to that village's position vis-à-vis Mount Bromo, and are further complicated by the fact that not all villagers recognize the older directional schema anymore but use the modern cardinal system.[26]

Let us dwell for a moment on the complexity of the cardinal and the indigenous (volcanic) systems of orientation intersecting. The former is a tool of cartographic governmentality[27] that surveys selectively, in values of x and y. The latter, by insisting on altitude, implicitly acknowledges human-environmental dynamics as conditioned by topographic and atmospheric variation. The cardinal system stops short of registering multidimensional relations and thus is unable to provide the coordinates in which they can evolve. Incapable of accessing the very ground of endogenous technics, Cartesian geography has been complicit in their erasure by design. To the extent that they have survived, endogenous technics have evolved *off the map*. Having evaded cartographic inscription, indigenous practices through which endogenous technics are expressed have been classified by the colonialists as primitive and underdeveloped, or "informal." Nonetheless, parallel to colonialist efforts to render landscape legible from a fixed perspective of cartographic norms, local livelihoods and culturally significant practices have continued to register and harness the ground rippling underneath, below (and thus beyond) the projection grid.

In 1855, Mount Bromo was etched into what is, problematically, still referred to as "the first accurate map of the island of Java."[28] In preparation of an external attempt to access the volcanic territory in order to tap into its productive capacity, this document charts the island's topographical spine as the site for the second phase of Dutch colonization.[29] Shifting the focus from the coast onto the island's interior where "everything which [could] be produced profitably . . . and have trade potential"[30] can be cultivated, the map provides the scheme for the design and construction of an infrastructure required by an entirely new extractive regime that would come to dominate the next 100 years of colonization. In 15 augmented views that zoom in on the volcanic highlands, the map acknowledges the origins of the island's wealth in the active geologic formations that capture the monsoon and produce rich soil. Made by the German-Dutch botanist and geologist Franz Wilhelm Junghuhn, the document is substantiated by 15 years' worth of field observations that are condensed into the four-volume report *Java's Shape, Flora, and Internal Structure* (1852–1854). An accompanying *Atlas of Views*[31] presents 11 bucolic scenes of picturesque landscapes, advertising to prospective European settlers idealized backdrops to investment and settlement: a new phase of the Dutch East Indian dream. Junghuhn's selective representations echo his position during the 19th-century debates on the acclimatization of European settlers in the tropics: "The Indies were a place where one could survive, live well, and even thrive, but only if one lived in the cooler mountainous regions, where the climate was indistinguishable from that of Europe."[32]

---

25   For an ethnographic perspective on the political ecology of the religious festival of Kasada, see Hefner, *Hindu Javanese*, 46–63.

26   Smith-Hefner, "Language and Social Identity," 9.

27   Cartographic governmentality relies on the flat and portable character of the map as a stabilizing agent in long-standing protocols that uphold centralized regimes via bureaucratic processes, as theorized by Foucault. Cardinal systems are the material substance of coloniality that runs through the governmental structure of postcolonial states. On the theoretical transition from postcolonialism to decoloniality, see Walter Mignolo, *Local Histories/Global Designs: Coloniality, Subaltern Knowledges, and Border Thinking* (Princeton: Princeton University Press, 2012). Alluding to this transition, Irmgard Emmelhainz suggests "dissolution of representation in favor of relation," in response to what she claims is the blind spot "habit" of coloniality: "The modern worlding of the world—which includes the production of objective reality by experimental science, knowledge, and design—coincides with the ruthless elimination and instrumentalization of certain creatures by others." See Irmgard Emmelhainz, "Fog or Smoke? Colonial Blindness and the Closure of Representation," *E-Flux Journal* 82 (May 2017): 5.

28   Alex Lehnerer and Philip Ursprung, "17 Volcanoes," lecture given at the Canadian Centre for Architecture, Montreal, September 29, 2016, https://youtu.be/jQ0Q7LQ4xTs (4:02).

29   The first phase was a corporate colony based on forced labor and the spice trade. Founded in 1602, the Dutch East India Company (VOC) is considered the first transnational corporation publicly trading in an official stock exchange. With an approximate shipment history of 5,000 voyages, the VOC held a global spice monopoly for most of the 17th century on nutmeg, mace, and cloves, extracted from the Maluku Islands in the Malay archipelago. To oversee an extended coastal urbanization through a network of fortified ports, by 1619 the VOC had based its overseas operations in the harbor city of Batavia (today Jakarta, Indonesia).

30   During the 1850s, liberal politics in the Netherlands dismantled the protectionism vowed to the Dutch East Indies, and a constitutional act granted the parliament power over the colonial assets for the first time. As a result, land taxes were replaced with a cultivation system which forced the indigenous population to plant 20 percent of their land with export crops and work shifts of 60 days per year in government-owned plantations, and Javanese land was opened to Dutch entrepreneurs. For a critical account of the role of politico-scientific bureaucracy in the expansionist project of the Dutch East Indies in the 19th century, see Andrew Goss, *The Floracrats: State-Sponsored Science and the Failure of the Enlightenment in Indonesia* (Madison: University of Wisconsin Press, 2011).

31   Franz Wilhelm Junghuhn, *Java: deszelfs gedaante, bekleeding en inwendige structuur* (Amsterdam: P. N. van Kampen, 1850–1854). The illustrated atlas (1854) is available online via the Biodiversity Heritage Library. See: http://dx.doi.org/10.5962/bhl.title.9503.

32   Henrik Pols, "Notes from Batavia, the Europeans' Graveyard: The 19th-Century Debate on Acclimatization in the Dutch East Indies," *Journal of the History of Medicine and Allied Sciences* 67, no. 1 (January 2012): 131.

Junghuhn's cartographic representation locks the island into a non-native, exogenous geographical imaginary. Along with the infrastructural interventions in the coastal and alluvial landscapes by the Dutch colonial administration, this cartographic reinterpretation would come to determine the trajectory of Java's urbanization that still persists today: a technocratic project hell-bent on "normalizing" deltas, rivers, ridges, and slopes in an attempt to channel the island's tropical metabolism into controllable flows.

By the mid-19th century, colonial policy-making and management were depending heavily on, and systematically deployed, scientific expeditions. In this context, Junghuhn was the ideal colonial bureaucrat-explorer. He had already drawn a cross-sectional scheme of the highlands as an altitudinal blueprint for real-estate promotion, which among other things made the establishment of the world's first pharmaceutical cartel possible: the introduction of the burgled Andean cinchona tree to the mountains of Java for the production of quinine, the antidote to malaria.[33] The implications of this development were no less than geopolitical in scale. If, as noted by historian Andrew Goss, "preventing the fever among soldiers and bureaucrats was a critical requirement for establishing European colonial rule in Africa and Asia,"[34] then cinchona promised a win-win model in which an in-demand cash crop was also the key remedy to promote, foster, and accelerate white settlements immunized against the very natural environments they occupied.

In the formation of scientific knowledge, theoretical articulations that fail to become institutionalized carry an experimental line of progression that reveals "forgotten frames of mind in their own terms."[35] These "errors and endings," Peter Galison argues, shift our gaze to certain factors beyond the original problem: "In discussing experiments we look for historical 'mistakes' just as geologists search for indicator minerals: in both cases as clues suggesting deeper-lying forces." Suspicious of the hegemonic theory of plate tectonics as the explanation of the origin and progression of "virtually all of the present and most ancient mountain belts on Earth,"[36] Batavia-born Dutch geologist Reinout Willem van Bemmelen was an active voice in the early-20th-century debate that split tectonics into a paradigmatic binary: *fixism* versus *mobilism*.

Van Bemmelen was the last chief of the Volcanological Survey of the Netherlands East Indies and chair of Economic Geology at Utrecht University until 1969. His mandate was to support accelerated extraction in the Dutch East Indies in the last stage of that colonial project, but his dissenting position within the discipline of geology is worthy of special attention. Between 1930 and 1977, his professional development was anchored in the vast geotectonic variations of the Indonesian archipelago. Characterizing the geology of this "island empire" as the avant-garde landscape for interrelated patterns of volcanic activity, seismicity, and gravity anomalies, van Bemmelen and his colleagues used it to illustrate and speculate on the principles of mountain building, or orogenesis. The Indonesian archipelago was cast as a "testing ground" for "the fundamental problems of the origin and evolution of our planet," due to an "orogenesis in full swing [where] the general relations between this phenomena can be better studied than in more or less completed mountain systems as the European Alps."[37]

Eventually van Bemmelen would become the first to claim that volcanoes, evolving in irregular intervals, are manifestations of forms of mountain building. This landscape formation would nourish van Bemmelen's skepticism toward the rising influence of plate tectonics theory, leading him to state: "Earth scientists should beware of 'rigid' dogmas, and strive for a 'plastic' adaptation of their concepts to the incessant flow of new, pertinent facts."

Born and fed on volcanic ash, van Bemmelen offered perceptions of orogeny that closely pursue the material dynamics of his home region. Differentiating between volcanic activity as chemical, and tectonics as mechanical transfers of energy from the core to the earth's surface, he considered plate tectonics an acceptable yet simplified synthesis of phenomena that should remain open to exploration. Emphasizing the impossibility for models to reproduce reality, his writing proposes alternative representations, using large-scale sections across volcanic grounds where "indirect circumstantial evidence plays a major role in the shaping of our views." This plastic notion of inference acknowledges the reality faced by volcanologists who search tephra for clues, while the volcano itself thwarts all expectations of finite evidence in support of an incontrovertible truth.

---

33  In 1851 Charles F. Pahud, then minister of colonies (later governor-general) of the Dutch East Indies, commissioned botanist Justus Carl Hasskarl to follow Humboldt's layouts of the Andes and traffic out of South America specimens and seeds of the cinchona tree. While the market for cinchona's bark—from which quinine alkaloid was extracted for medical use against malaria—was controlled by the newly independent nations of Bolivia and Peru, its commercialization had become significant to the expansion of European empires. Java's highlands seemed suited to sustain another supply store. Appointed inspector of scientific research in 1856, Junghuhn directed the plantation and propagation of more than one million *Cinchona pahudiana* trees until his death in 1864. This expanded the colonial frontier and allowed a new cohort of "floracrats" to drive Java's production of quinine up to 80 percent of the world's supply by 1900. See Goss, *The Floracrats*, 56.

34  Goss, *The Floracrats*, 35.

35  Peter Galison, *How Experiments End* (Chicago: University of Chicago Press, 1987), 5.

36  Wolfgang Frisch, Martin Meschede, and Ronald Blakey, "Plate Tectonics and Mountain Building," in *Plate Tectonics: Continental Drift and Mountain Building* (Berlin: Springer, 2011), 149–58.

37  Reinout Willem van Bemmelen, *Mountain Building: A Study Primarily Based on Indonesia, Region of the World's Most Active Crustal Deformations* (The Hague: Martinus Nijhoff, 1954). Coincidentally, Junghuhn also previously compared life in the Tengger territory to that of the Swiss Alps: "They do such European things as building their houses with planks (not atap or tropical thatch) and keeping a fire going in a hearth. Their homes are communal, theft is unknown, and they keep herds of buffalos which return voluntarily to their stables at night, as if they grazed in the Swiss Alps." See E. M. Beekman, "Junghuhn's Perception of Javanese Nature," *Canadian Journal of Netherlandic Studies / Revue canadienne d'études néerlandaises* 12, no. 1 (1991): 17–18.

# Enter the Blanks

Conceptualizing volcanic ejecta as a cumulative process, the manifestation of which every now and then expands or rewrites the principles of an established theory, eruptions are represented in van Bemmelen's drawings as a series of blanks. In this regard, his cross-sections, connecting craters and calderas as thresholds to an inaccessible planetary interior, parallel those of Kircher's *Mundus Subterraneus*. Magma chambers and subduction zones continue to evade accurate graphic representation. An early-19th-century classroom chart, *Sectional View of the Crust of the Earth* (1828–1848) by Orra White Hitchcock, presents this suggestive power of voided, blank space at a vast, planetary scale.

If scrutiny of the volcanic allowed van Bemmelen to question and venture beyond the rising authority of plate tectonics as a universal theory, then the volcanic itself, in our analysis, marks one of those diagnostic frameworks that enable us to challenge a number of contemporary strategies that have evolved out of coloniality: among others, the spatial politics of environmental risk mitigation. Over millennia, a range of sophisticated technics and patterns of human habitation have negotiated landscape contingency on its regional, idiosyncratic premises—technics and patterns that have persisted and adapted alongside urbanization models built on Cartesian geography. Challenging the logic of containment and fixity of form, the volcanic sensitizes us to the long-standing existence of dynamic and mobile configurations of human and nonhuman producers of space. Most important, it dares us to rearticulate what design that pursues such productive entanglements might be, equally embedded in specific local conditions and planetary processes.

The concept of geotechnicity, while speculative at the outset, opens an investigation of prospective fields of study and design for more-than-human cohabitation. It provides a diagnostic framework for retracing the ways in which multiple nonhuman agencies have contributed, and continue to compete, in the propagation of matter and the production of space. "Geographies of the posthuman," then, names a new phase in the practices of recording, reading, and relating to landscape as a web of local symptoms whose origins are distributed and interconnected in the stratigraphy of the planetary body. It obliges us to locate and distinguish between manifestations of geotechnicity and those of bounded technological systems, respectively, as materializing imprints of two distinct machines. It helps us identify the relative position of any given population vis-à-vis the rift between the functioning of these machines. It leads us to reconceptualize our agency as a question of implication in the momentum of either, as the crux of our survival, of our redesign.

**Image Credits**
157: Digital archives of Stanford University. www.web.stanford.edu/group/kircher/cgi-bin/site/?attachment_id=593

159: Archives & Special Collections at Amherst College. www.acdc.amherst.edu/view/asc:20017. In the public domain.

165: This investigation has been visually explored with in collaboration with landscape architect Andrew Boyd. Image courtesy of Andrew Boyd.

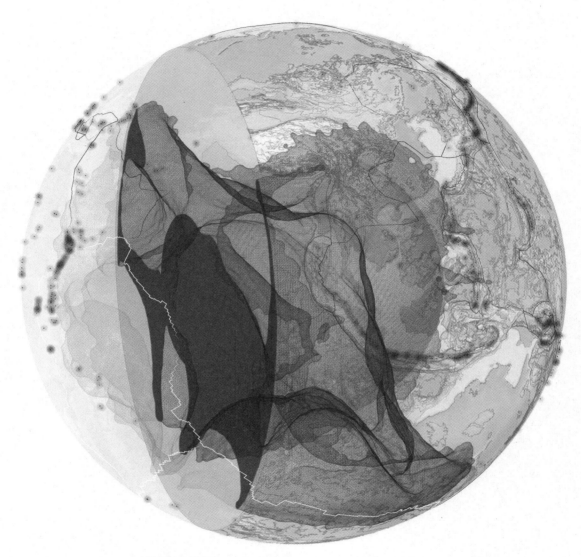

**Tectonics: Mantle Convection as a unified engine of geologic renewal, a substrate that produces the largest and densest settlements in human history.**

Namik Mackic & Pedro Aparicio Llorente

Volcano, East Java, 2016.

We coexist with human lifeforms, nonhuman lifeforms, and non-lifeforms, on the insides of a series of gigantic entities with whom we also coexist: the ecosystem, biosphere, climate, planet, Solar System. A multiple series of nested Russian dolls. Whales within whales within whales. . . . When we look for the environment, what we find are discrete lifeforms, non-life, and their relationships. But no matter how hard we look, we won't find a container in which they all fit; in particular we won't find an umbrella that unifies them, such as world, environment, ecosystem, or even, astonishingly, Earth.
— Timothy Morton, 2013

# Critical Ecologies of Posthumanism

# An Interview with Cary Wolfe

| | |
|---|---|
| New Geographies | In *What Is Posthumanism?* (2009), you claim that the "human" occupies a new place in the world, among a larger constellation of other life-forms and "nonhuman subjects," each with their own "temporalities, perceptual modalities, and their own forms of environment." On this last point, specifically, how would you articulate an understanding of the *environment* from a posthumanist perspective? And how would this notion challenge the (humanist) logic of the production of environmental space as a differential field in which humans appropriate, design, and administer—often through institutionalized techniques of exclusion and subjugation—the space of the nonhuman? |
| Cary Wolfe | For me, a posthumanist understanding of the place of the human in the world is both new and old—in this sense, "post-" does not simply mark a chronological succession in which, for example, posthumanism comes after humanism. This is not simply a matter of the old schema of temporality as a site of overlapping "dominant," "residual," and "emergent" tendencies, but something far more radical and paradoxical: a "strange loop" (as I called it in my second book, *Critical Environments*, 1998) in which the past, the present, and the future are caught in a ceaseless reconjugation and redefinition of each other. To put it in slightly more familiar terms, the sense here is genealogical rather than historical (touching on Michel Foucault's classic essay "Nietzsche, Genealogy, History" [1971] which also makes it clear why a dialectical schema is not adequate to the context either). In any case, a posthumanist understanding of the place of the human is "new" in that it draws upon contemporary philosophical and theoretical resources that challenge humanist anthropocentrism and show how the radically de-centered human is enmeshed in relation to its environment and the other life-forms that populate it: a fact that is dramatized at the current historical moment of the Anthropocene, characterized by global warming, species loss, and so on. But it is "old" in the sense that those very same resources allow us to see that the human has never been "master in its own house," so to speak, that it has always been the product of a multitude of relations and forces that are not properly its own—and not even living, or organic. Thus posthumanism actually sees the human as *coming before* the domestication of that heterogeneous set of forces and relations in the historical development that we now call "humanism." (I'll simply note here that I am using shorthand, because humanism is obviously a huge, rich, and diverse area on its own.) This means that the posthumanist perspective on the human is always already radically nonhuman (or better still: ahuman) down to its very core (as so much recent work on the microbiome has shown in quite literal terms), and it also means that the term "environment" cannot be understood as an autonomous entity antecedent to any particular set of relations, but as something that is always part of a dynamic, enfolded system-environment relationship. To the microflora in your gut biome—without which "you" could not function as a "human"—"you" are the environment. So the outside *is* the inside—but it's always a particular, nongeneric environment that is enfolded in highly specific ways; a necessity, from a systems theory point of view, since any system has to cope with an environment that is, by definition, exponentially more complex than the system itself. We do indeed have to administer the environment in the old-fashioned sense; this just helps explain why the blind spots created by the tautological self-reference of our own organism-environment relationship create so many "unintended byproducts" (to use Kenneth Burke's wonderful phrase), even when we're trying to be sensitive and responsible about what we're doing. This is not a problem that one can simply make go away—either by appeals to immanence or transcendence. Indeed, ethical and political responsibility begin with the fact that there *is no* given ground or foundation from which to work, from which the inescapable blindness and partiality of our own situatedness might be surmounted. |

**NG** The perception of the human as developed in the Western humanist tradition—perhaps most clearly symbolized by Leonardo da Vinci's *Vitruvian man* (c. 1490)—leaves the meaning of its opposite, "otherness," unquestioned in all its manifold expressions: marginalized races and ethnicities, genders and sexualities other than straight white male, animals, ecosystems, and so on. Various critics of the humanist tradition have historically fought against representations of others that portray them as less-than-human, or render them as disposable beings whose lives hold less value. Given this trajectory, how do we instead use a spatial-analytical lens to extend posthumanist criticism to include new frameworks for understanding the geographies and territories of others, whether human or nonhuman? Is it fair to start from peripheral and marginal spaces far from the centers of power and political imagination—in lands that are marked as exploitable resources, places to be forfeited, contaminated, and forgotten; or, at the other extreme, from sites meant to be left "untouched," protected, and preserved like national parks and conservation landscapes? How can we apply posthumanism to these other spaces as part of a larger critique of the Western humanist tradition? Or should these spaces rather be considered as a form of otherness itself, that is, the other side of the spatial project of progress and technological development?

**CW** To be sure, otherness has to be rethought outside of and in resistance to the discursive technologies and cartographies that have traditionally marked the other (whether human or nonhuman) as "killable but not murderable"—typically through racialization, and animalization, but not just that. But as you will have already guessed from my previous response, I do not think that an antidote to this machinery of violence involves a kind of benevolent recovery mission in which we see otherness more clearly and value it for what it is, even though I'm sympathetic with the impulse to do so. Rather, the antidote involves a more profound, thoroughgoing deconstruction of the very human who does the valuing, and who conducts the recovery mission. After all (as I suggested in some of my earlier work), this became a problem in so-called animal rights philosophy, despite its admirable anti-anthropocentrism: it wanted to value the moral standing of the nonhuman other more explicitly, but left intact the humanist philosophical schema of subjectivity regarding the evaluator. Truly posthumanist work requires the development of a theoretical and philosophical approach to these problems that is "heterogeneous" (as Jacques Derrida puts it) to the logic of self and other. How is this related to the point in my prior response about self-reference and autopoiesis? It is precisely the inescapable self-reference of any autopoietic system that creates the "blind spot" (Niklas Luhmann's term) which calls for other points of view on the world, whether human or nonhuman. But of course, the same condition obtains for the other—regarding its own inescapable blind spot. If everything I've said so far makes sense—that there is no given environment (what we used to call "nature") that is antecedent to any specific set of relations—then there is hardly a generic answer to your question about where to begin.

Thinking this heterogeneous logic is only half the problem; the philosophical and theoretical half, and in many ways the easy half, but also the absolutely essential half—a precondition to any concept of the political whatsoever. But the other half of the problem is biopolitical: that is, the uneven distribution of the material effects of this blindness on particular populations of living beings, human and nonhuman. As your question suggests, in the industrialized West "normal everyday life" is absolutely predicated upon mass, routinized violence toward the nonhuman world, and in ways such that some beings, human and nonhuman, are more caught up as victims than others, just as some human populations are more culpable than others for producing this violence: not only through directed killing of the sort we see in industrialized food production, for example, but also through the "terrorism of letting die" that Derrida describes in *Philosophy in a Time of Terror* (2003). This indifference is not so easy to distinguish from the fundamentalist terrorism that we in the West feel to be so morally beneath us. But who is this "we," after all? Both biopolitically and anthropologically, I am very reluctant to generalize about human beings. But in this context, one psychic function of "untouched" nature that sustains the fantasy of the national park (as Margret Grebowicz suggests in her neat little book, *The National Park to Come*, 2015) is to produce the idea that "we" are all equal, one people in the eyes of a nature that predates our social institutions. I need not belabor the point that such an idea is counterfactual; national parks are heavily managed and technologized landscapes. This does not mean there are no Pleistocene-era creatures living there that could kill you, nor does it mean there aren't a hundred ways to die in what remain, for all that, wilderness areas. To put it another way—and I touched on this in my piece for the previous issue of *New Geographies*—the reality of these landscapes is much more complex than the work of fantasy that they enable in the psychic realm.

NG  In rethinking the humanist nature-culture dichotomy, you discuss the entanglements between "green ecologies" of the living, the organic, and the biomass; and "gray ecologies" of the machinic, the technological, and the electronic. Much of *What Is Posthumanism?* examines the problem of the nonhuman as a complex constellation of different life-forms (the animal, the environment and its ecosystems, etc.), that is to say, as part of the diverse lineage of so-called green ecologies. But what about the nonhuman as a "technological other"—as part of the artificial lineage of gray ecologies: abstract and material systems such as capital, financial markets, or technological networks created by humans, which in their current degree of development surpass or extend well beyond human agency and control? How do these technological others further complicate human-nonhuman relations, from a posthumanist perspective?

CW  You are right that my earlier work on posthumanism focused on green ecologies and, even more specifically, the animal question. But from the very beginning—and for the very reasons that animate your question—I have always insisted that "animal studies" and the question of the nonhuman other are but a subset of the much larger problematic of posthumanism (which is one reason that the *Posthumanities* series that I started in 2007 at the University of Minnesota Press is not an "animal studies" series!). The "gray" ecologies that you mention are, historically speaking, contemporary instances, but in fact what we call the "human" has always been, in its very origins and evolution, a technologized and prosthetic being, whether we are talking about tool use, language, archives and storage technologies of various kinds, and so on, as thinkers from André Leroi-Gourhan to Derrida, Bernard Stiegler, Gregory Bateson and many others have argued. This is directly related to my point about the human not being "master in its own house." As all of these thinkers have shown—but we could name many more, of course—the human only becomes human by being radically subjected to something it is not, something external that was on the scene before the discrete individual arrived, and that will be on the scene after that individual is gone. Freud had his own version of this concept in *Civilization and Its Discontents* (1930), but Derrida's rendering is, I think, especially powerful and wide-ranging. He notes that the *machinalité* of semiotic systems of all kinds has a "spectralizing" effect on those who use them, because the systems' functionality is not dependent upon the existence of any particular empirical being. Thus, the logic of the trace by which such systems operate (logics that are heterogeneous to concepts of "self" and "other," as I argued above) traverses and unsettles not only the life-death relation but more radically still, the machinic-organic relation. (As Bateson says, Socrates is still alive, more alive than ever, even though the concrete bioenergetic individual named "Socrates" is long dead.) As Derrida points out, the philosophical tradition has typically attempted to stabilize and domesticate this untidy state of affairs by insisting on the purity of the opposition between genuine "response" (associated with humans) and merely mechanical or automatic "reaction" (associated with animals). But of course, the purity of that opposition cannot be maintained, as he demonstrates, and that's one of the reasons (but only one) that I have never found the concept of "agency" very useful in addressing these questions—or the contemporary developments and challenges that your question rightly foregrounds.

NG — An important chapter of your book concentrates on disciplinarity and the underlying assumptions of the humanist disciplines, which have, especially in the second half of the 20th century, expanded their fields of inquiry to incorporate pressing problems and so-called marginalized groups to their institutional apparatus, albeit without "destabilizing the schema of the human that undertakes such pluralization." We would like to foreground the design disciplines here—architecture, landscape, urbanism, environmental design—which have become increasingly conscious of their role in the process of spatial structuring and the domination of the planet as a terraformed totality. These processes of organization can produce ecological damage and environmental destruction as the "other" side of progress, endless growth, and technological evolution, quite often wrapped in the allure of aesthetics and innovation. In what way do you think design embodies a humanist form of disciplinarity, and what could be considered problematic about this according to a posthumanist ethics?

CW — I'm probably not as equipped to answer that question as you are! I got into this a little bit in the chapters on art and architecture, where the key terms of engagement were representationalism and composition. I was—and still am—interested in the extent to which representationalism is inevitably a kind of speciesism. I am playing devil's advocate here, but my point is simple: representationalism—what counts as a representation—is always indexed to the particular sensorium, with its various capacities, scales of perception, and so on, of a perceiving being. (Leaving aside, for the moment, the second-order cultural conventions around representation that are in force here as well). A recognizable image is put before a viewer that in some way or another lines up with that viewer's world. Now one might draw a straight line here from the Renaissance theory of perspective to Freud on the eye and the upright stance of the human in *Civilization and Its Discontents*, to Sartre on the Gaze, to Foucault's reading of Jeremy Bentham's Panopticon, to Laura Mulvey and feminist film theory of the 1970s and 1980s—but my point is a little different. It's simply to remind us that there are other ways of being in the world where the relationship between the sensorium, the world, and the image (if that's the word we want here) is quite different. This includes other human beings, by the way, as my discussion of disability, visuality, and the case of Temple Grandin tried to make clear. So how does taking these other ways of being in the world seriously—that dogs and many other animals experience the world primarily through what they smell, that bats and dolphins use echolocation and seem to communicate in quasi-digital ways, just to take two dramatic examples—affect the strategies and media deployed in the work of art or design? I know artists who would insist that you could never make a work addressing these kinds of issues in the medium of painting, for example, in part because the relationship between medium, representation, and sensorium (i.e., the dominant visual sense associated with the human) is so overdetermined. Many of the same questions are in play in the fundamentally compositional approaches of design and architecture. I was attracted to Rem Koolhaas and Bruce Mau's *Tree City* (2000) project for Downsview Park in Toronto because of what they referred to as (a very funny phrase) their "refusal of the realm officially known as architecture." For them, it was a refusal of over-programming (some would say "any"); a refusal to build buildings and thereby compositionally fix the space. They preferred what systems theory would call a "loose coupling" of the space and the activities of those using it, allowing them to evolve together over time. So in an interesting way, time rather than space became the primary architectural medium. By contrast, the compositional impulse attempts to freeze or at least severely constrain temporality and the work of time on space. In deconstructionist terms, Koolhaas and Mau were up to something like a temporalization of space, a "becoming-time of space." The design question for *Tree City* was not "Does it work?" in the sense of successful programming, but rather, as I put it to Mau's great amusement: "Can you wait?"

NG — Posthumanism, as you conceptualize it, undermines social and interspecies hierarchical structures and advocates for cohabitation in a "shared environment" produced by the recursive actions of beings coexisting within a "consensual domain" (to quote Luhmann). Parallel to this, the notion of continuity and integration has been developed from a historical perspective since Word War II, which witnessed the formation of a persistent ideology of unity and planetary governance manifested in the establishment of world organizations, global trade agreements, and transnational infrastructural networks. However, at our current historical conjuncture, we face a situation in which the world is increasingly socially, politically, and ideologically polarized, segmented by strong borders, walls, and barriers. The Brexit vote in the UK and the rise of white nationalism in Europe, Donald Trump's administration in the US, the return of ultraconservative governments in Latin America, and endless military conflict in the Middle East are all examples of these global trends. Is it fair to say these territorial and ideological fragmentations signify the deep crisis that the Western humanist project is currently undergoing? If this is the case, how can a posthumanist approach elucidate—or provide an alternative for—the complexities of an increasingly divided world?

CW  I'd be reluctant to draw very direct lines between the phenomena you describe and the legacy of Western humanism. Is Donald Trump a humanist? No. I would say he is "not even" or "not yet" a humanist. And there are many strains of humanism—for example, secular humanism—that would be opposed to polarization, discrimination, and violence. Be that as it may, one way to begin to answer your question is to reiterate that, for me, the political is about constitutive social antagonism, which is rooted in the inescapable self-reference or partiality of our situatedness. This is not exactly the same as the usual ideas of pluralism or tolerance, because what necessitates the place of the other is structural, rather than psychological—that is, it is not simply a matter of "having a good attitude" or taking thought, but rather of an alterity built into the very core of any identity; its constitutive blindnesses. From this vantage, ideas such as consensus and community, while understandable, are of little use. The problem is not constitutive social antagonism itself. The problem is that it is overlaid—I'll limit myself to the US context here—by a political system that disproportionately represents conservative and corporate interests, and that incentivizes demagoguery and ideological polarization rather than pragmatism, problem solving, and compromise. I'll come back to this in a moment. But to make your question a bit more manageable, I'll frame it in biopolitical terms, the chief coordinates of which, for your question, would be these: What is the relationship between the refined mechanisms of "governmentality" (to use Foucault's language) and the continued exclusion of various populations from what Roberto Esposito, Derrida, and others call—after Donna Haraway and Luhmann—"immunitary" protection, as members of a world community? This question is not related in any simple way to what we know as "the political," nor is it entirely new, as Hannah Arendt's post-World War II rumination on "the right to have rights" makes clear. In fact, I think Foucault is right that we find a qualitative shift in the very logic of the political as we move from sovereignty to disciplinarity to governmentality. (Not that any prior mode ever goes away completely, as the exercise of executive power in the "state of exception" evoked by both the Bush and Obama administrations indicates.) Foucault recognizes (as does Luhmann, and both Deleuze and Negri in their reflections on the "control society") that the political is situated in a context of increasing social complexity in which no single system, including the political, can steer or dictate in any direct way the operations of the other social instances (most strikingly, of course, the economic—witness QE2 and QE3 in the US). Hence Foucault's interest in neoliberalism in his lectures at the Collège de France in 1978–1979, translated into English as *The Birth of Biopolitics* (2004). Under such conditions, a posthumanist understanding of the political would insist that you have to have a requisite theory of social complexity *before* you can presume to know what "the political" is or what its modes of effectivity might be. I think this is where Foucault's interest in neoliberalism comes from. In my view, the semantics that dominate political philosophy, and the thinking of some of our most prominent political philosophers, are woefully inadequate to this task.

My own feeling is that what we see now is an intensification of something that has always obtained for the political system in its modern form: that it exists to reproduce itself, and will take on what burdens it must to accomplish that task, in the name of what Habermas famously called "legitimation." To me, the primary problems with the political system in the West are two: first, at least in the US, the political system is corrupt—essentially corrupt, and as far as I can see, irreparably corrupt. This was the case long before the *Citizens United v. Federal Election Commission* (2010) ruling in the Supreme Court, which allowed for unlimited corporate spending on electioneering communications in the name of "free speech." The only people who can remedy this corruption are the very people who profit from it. This doesn't mean that all elected officials are cynical shills—many are high-minded, I am sure. It simply means that the system works the way it works. This is not a philosophical problem, as if more or better philosophy would fix it. In fact, I would be happy to see a series of mundane, "non-ideological" (so to speak) reforms enacted—focused on eliminating gerrymandering of congressional districts (currently under review by the Supreme Court), greater voter representation of the actual heterogeneity of the adult population, eradicating corporate money from the political system, and so on. The results would be quite dramatic.

The second primary problem with the political system in the West—and this is not just a problem for the US, as was dramatized by Brexit, as you point out—is that roughly half the voting electorate thinks the other half is crazy. Again, this has always been the case, but it has greatly intensified since the 1980s because of changes in the so-called public sphere linked, not least, to the changing role of corporate-owned mass media, at least in the US. So any serious political discussion has to begin with this question: what is the relationship of the political system to that fact? In this light the self-referentiality and, in a sense, the diminution of the political vis-à-vis the greater context of social complexity can be seen as a good thing or a bad thing. When George W. Bush was in power, it seemed good to me, because the autopoiesis of the political system short-circuited any attempt by "W" to unilaterally impose his will via the political system. But of course, when Obama was in power, the very same autopoiesis was immensely frustrating. I'll conclude here—because I've already gone on long enough, and because I think we can find here a posthumanist way of thinking anew, for better and for worse, the question of pluralism in an increasingly divided world. Perhaps the role of the "political" as "representative" of this heterogeneity and plurality takes on, under this understanding, a diminished form, both conceptually and pragmatically. And perhaps that is not such a bad thing.

# Adventures in Third Nature

**McKenzie Wark**

We're at a campground on Mount Zion, in the American Southwest. We are here to see nature. They have great nature here. Mount Zion is a national park whose main attraction is a particular kind of natural sublime. The vast canyons and other rock formations impress themselves on you and take your breath away. You are left momentarily speechless—but only momentarily. Then, you can work your way back to some sort of reconciliation with the great outdoors. And take a selfie in it.

It would be easy to satirize the antics of today's nature tourists, although if I am being honest I would have to admit that I, too, took selfies with my little family group of holiday campers. What might be more interesting would be to try to parse the layers of relation to nature involved here. And the campground seems a good place to start, having already observed some of its contemporary rituals, of which a selfie-in-the-wild is just one.

We are in what Americans call a recreational vehicle, or RV. It is basically a little hotel room on wheels, with a refrigerator and stove that run on propane gas, and a generator that can power the air conditioner and the lights. There is a shower, and a toilet with a tank for fresh water and a tank for waste. Here there is a full-service hookup, where electricity and running water are provided, as is an outlet for waste, so we can join our little capsule to the mains for now.

What we have, with our RV, is a sort of miniature pod version of a built environment, which twists and folds and channels nature into forms amenable to comfortable human life. Perhaps we could call this a "second nature." It is made out of nature, out of natural properties like water and gas and electricity and of course the gasoline that powers the engine to travel about. After a while, this second nature feels quite natural. A few days into our trip we were cooking on the RV's little stove with food from its little fridge, using its toilet, and running its air conditioner against the blazing heat of summer in the American Southwest.

I started wondering if campground culture might be more about this second nature than about nature. I don't know much about nature, but I am from an industrial town and from a long line of engineers, so I am used to thinking about how stuff is made, how it works, and who made it. In some ways, nature appears as a bit of an alibi for experiencing this variant second nature. People leave their regular homes, their routine version of second nature. They rent an RV or bring the one that sits in the driveway for most of the year, and experience this *different* second nature as a way of achieving some distance from their everyday one.

What got me thinking about this was watching a new neighbor pull in to the campsite next to us. What appeared to be the family patriarch took charge, directing everyone in setting up the site. Their rather long caravan had jacks that could be extended from its corners to the ground to stabilize the entire structure. These were like regular car jacks for changing tires, except that they extend downward from the body of the vehicle, to which they are attached, rather than up from the ground. The man took an electric drill out of his pickup truck, which he had thoughtfully brought along to crank the jacks down. Afterwards he put it back into his pickup, which had the Ford logo emblazoned on its grill.

I do not know much about which companies make the kind of second nature machinery and gear most favored by campers, so I took a look around. Besides Ford pickups, I recognized the Winnebago brand, as they are famous in the United States. Winnebago initiated the mass production of RVs in 1966. There was also a single, gleaming, silver Airstream trailer, which I learned via Google is rather expensive, and bit of a cult item. Its streamlined, polished aluminum look dates from 1936. Basically, it seems there are entire manufacturing industries that make stuff specifically so that one can spend the summer break or a long weekend driving to nature in order to enjoy this kind of adapted second nature, which provides many of the comforts of one's regular second nature, or home. I even saw a satellite dish.

What I want to draw attention to here, first, is the possibility that the experience of camping is less about leaving second nature to achieve some appreciation of nature than about leaving second nature to experience another second nature. Perhaps this was always so. To even form a concept of nature, one might need to inhabit some fold within it that creates some distance and that makes nature appear as something that can be conceptually understood. For us moderns, for whom God is dead, that other inhabitation might have to be a second nature, or, as I shall attempt to demonstrate shortly, a third.

This might be the case no matter how far one ventures into nature and away from the comforts of second nature. I'm thinking of the Werner Herzog movie *Grizzy Man* (2005), about a young man named Timothy Treadwell who decided to live with grizzly bears. It is not much of a spoiler alert to point out that a bear eats him. For Herzog, the film is about dispelling a notion of nature as bountiful and beautiful and with which the human can be in harmony. Nature is death and violence in a Herzog movie.

The thing I noticed, however, is that even Treadwell needed gear. He had state-of-the-art kit for camping, and video recorders, one of which was on when he died. In a moment of genius, Herzog holds the camera on Treadwell's girlfriend listening via headphones to only the audio of his death. We, the viewers, neither see nor hear it. Nature at its most sublime, beyond the fold of second nature, cannot be represented, and is death. To truly desire nature is to be at one with the death drive. And yet a complex mechanical and industrial second nature has to be there, at the edge of nature, at the lips of death, to even make it appear at all.

Something always gets in between us and nature. There is a second nature that intervenes. The experience of nature is via a second nature. Kant's sublime nature is not how a peasant sees the world. It is not even how a worker sees the world. It is how a bourgeois sees the world on his day off. He is not looking at a parcel of land for development on his day off. He is looking at the inverse of that. Perhaps it is the handiwork of God, or more likely, the work of its own inherent (lawful) processes of development. The bourgeois God is the God of Spinoza: so industrious, always busy, always in development. The bourgeois looks at nature as a mirror of the second nature he will command into existence—when he returns to the office on Monday.

It was a German Romantic kind of naturalism that gave us this busy Spinozan God, who comes to us via Johan Wolfgang von Goethe and Ernst Haeckel. Each thinker looked with curiosity at a nature sublime in its totality, but beautiful in its details and inner workings. Perhaps there is some rich cultural residue that brings so many German tourists, with their concepts of nature, on these camping encounters to Mount Zion. Maybe they are still searching for a nature of awe and reconciliation, unlike us more frivolous Anglophones who, after Burke, experience the sublime as the affect of delight. I don't really know.

What I did observe is a nature that could be experienced via the industrial culture of second nature. Basically, a nature you can drive to, and even drive through. Mount Zion has a road that winds all through it, including a tunnel completed in 1930 that is over a mile long. Since today's buses and RVs are so big, the traffic only flows in one direction at a time. There are slots cut into the mountainside that run perpendicular to the tunnel so you can see sublime subterranean nature from your car window, as you drive through it.

We were warned about traffic at Mount Zion. The RVs are slow. Even in regular automobiles, tourists slow down to take in the scenery. Meanwhile, the locals who know these twisty roads overtake at high speeds. For while sublime nature is an artifact of leisure, access to that nature is somebody else's work. What is rendered barely visible by the leisure of second nature is that second nature is also work.

So far I have mostly described some everyday experiences of second nature as it interfaces with a popular and mass-produced version of nature as sublime. I stressed its prehistory in European ideas about nature as sublime, but one could also add here more American ideas about wilderness as something to be conserved, as if it were a remnant of a biblical landscape. Indeed, many of the features of Zion National Park have biblical names, including the peaks known as the Three Patriarchs.

After Zion we traveled in our little bubble of second nature—that is, the RV—to Bryce Canyon. Descending into it and climbing back out of it on the back of a mule certainly helps understand how second nature has developed over time. The famous outlaw Butch Cassidy used Bryce Canyon as a hideout. In local oral history, Butch lived a long life around here and is buried nearby. Descending into the canyon on a mule, one can't help but think that challenges from the terrain, and keeping food in store, might have been greater obstacles to the outlaw life than the law. Indigenous people kept away from Bryce Canyon for spiritual reasons, which seem, as such reasons often do, to also be practical reasons.

When looking at all this sublime nature, it is hard to keep in mind that a vast infrastructure of second nature makes it possible to do so. Sometimes it seems it is just us and nature, but all around us are systems keeping flows of standardized commodities clicking along to meet our needs. There is water running out of taps, there is propane gas, there are places to get food, and of course, gas stations to keep the vehicle running. You can even get precut sticks on which to roast marshmallows.

On the other hand, it gets harder and harder to hide. After the famous horseback-and-pistol outlaws of the Hollywood westerns, there were the car-and-machine-gun outlaws of the early 20th century, but each were a dying breed. Telegraphy and radio made it difficult to outwit the law. The matrix of second nature brings the legal system with it, such as it is. Ben Morea, of the radical street gang Up Against The Wall Motherfuckers, once told me that when he had some trouble with the law in New York City, he hid out in New Mexico for a while, not really intending to return. But that was the 1970s. If you could still escape in those days, second nature has since grown in intensity and density, and now it is close to impossible to disappear in the era of third nature.

After leaving Bryce Canyon we stopped in a remote campground in the mountains, cool and green. There was no hookup, so we had to run the RV's generator for a bit to power the lighting. For a moment it was just us and nature, until a park ranger happened by to check whether we had paid our 12 dollars for the site. There were envelopes for the fee and a box with drop-slot to accept the deposit. I had to borrow money from my son, as we were nearly out of cash. He was disappointed that we weren't in cellphone range, but thanks to the generator, there was power to run his iPad.

If second nature is something built by collective human labor to make a more habitable nature, *third nature* is something built by collective human endeavor to try to overcome the shortcomings of second nature. I already hinted at some of its precursor forms. Telegraphy and radio move information much faster than you can move, say, an outlaw band on horseback or in a getaway car. Both are instances of what I call the vector. By vector, I just mean a line of fixed length but of no fixed position, like in simple plane geometry. It has one fixed aspect: that it is a line of given length. And one unfixed: it can be in any position whatsoever. It seems like a useful tool for building a concept about how a third nature might be constructed. A vector is made of information, but that information needs its own infrastructure.

Driving around the Southwest, one passes straight rows of telephone transmission towers that stretch to the horizon. Telegraphy and telephony do not have as glamorous a story as the railways, that iconic early industrial form of second nature. But they made the railways possible. In the classic western *High Noon* (1952), everything hinges on the train arriving in Gary Cooper's New Mexico Territory town for the outlaws to rob but also on preventing word of the ambush getting out in advance by telegraphy.

The landscape of *High Noon* is that of farming and ranching organized over a vast territory. How could the buying and selling of product across such a region take place? As James Carey proposed in a famous essay from his book *Communication as Culture* (1989), the modern sense of market is derived from telegraphy. The telegraph enabled the transmission of signals not just about train movements—or the movement of outlaws—but also prices. In the language I am using here, the term "market" loses its geographical specificity as a place people come to discover prices on the terrain of a second nature. Rather, it starts to name a feature of a more abstract space, a third nature.

Telegraphy is the start of third nature, of a terrain abstracted from second nature, where information can move more freely and quickly than either people or things. But it is just barely that. As industrial second nature expanded through the railways and the Fordist era of the mass-produced automobile, third nature struggled to keep up. The telegraph gave way to telephony and radio. These are developments of what I will call the *extensive* vector. What also had to develop was the intensive vector, firstly through systems of administration, but most thoroughly through systems of computation. Where the extensive vector wraps nature and second nature in its wires and radio waves to move information faster than people or things, the intensive vector stores and processes that information, faster than people or things.

Some of the most significant technical advances of the 20th century were developments of the intensive and extensive vector, and several were derived from trying to send not just Morse code but regular human voices over electrical wires, from one side of the American continent to the other. Telegraphy gradually made America a single market, but having to rely on skilled professionals to tap out messages and decode them at the other end of the line slowed things down. But getting audible voice communication to happen across a continental divide is a lot harder. Unlike in Europe, the United States did not approach the problem through a government body, but with a regulated private monopoly, one arm of which, Bell Laboratories, was charged with starting what we now know as information science.

A few miles from Bryce Canyon is a place called Kodachrome Basin. Photographers from *National Geographic* named it in 1948, after a film stock that Kodak first put on the market in 1935. Kodachrome was discontinued in 2009. These days, almost everyone uses digital cameras, perhaps built into their phones, as I do. I uploaded a picture of Kodachrome Basin to Facebook, as it had given me the idea for this essay. Photography was one of several mechanical and chemical extrusions out of second nature that were attempts to produce a third.

In Walter Benjamin's famous essay "The Work of Art in the Age of Mechanical Reproduction" (1936), *mechanical reproducibility* is supposed to make the world of organized industrial labor—that is, second nature—visible and intelligible to itself. But it never quite worked out that way. The perceptual and organizational forms extruded from industrial second nature made images of nature using what now seem like bulky and involved mechanical and chemical processes. It had a rather involved infrastructure. I can remember when, if you shot on Kodachrome, you had to mail the 35mm film canister to Kodak for processing, and they sent you back the slides. Third nature has done away with all that. Now I just snap a view with my phone, upload it over the cellular telephony vector, and all my friends can see it immediately.

Now it is time to confess that as my family and I travel around the Southwest, producing experiences of nature molded by the second nature of our well-equipped RV, there is something else in play as well. The whole excursion was planned by my partner using information found via Google. Various RV rental companies and campsites were assessed by their Yelp reviews. Plane tickets and site reservations were booked online with a credit card. Advice on what to see was sought over Facebook from various friends. The weather was checked on an iPhone app. What makes this experience of nature—in fact experienced via second nature—something that could be marshaled and executed is: third nature.

An electric drill in the hands of a camper is not the whole of second nature; an iPhone in the hands of a family vacation planner is not the whole of third nature. Both appear as if they are manifest for our convenience, and make the world appear as if it is manifest for our convenience. However, what we experience in both cases is only a part of the whole. And the part and whole fit together, in both cases, in rather different ways. The electric drill can wind down the jack under this particular camper van, right here; the iPhone can connect its user to information about this campsite, or another, or a Wikipedia page on the Winnebago, or a Netflix video for my teenage son who is rather bored with the whole family RV adventure.

Second nature makes nature appear as its double, in its image. It makes nature appear, per Heidegger's concept, as a "standing-reserve." Third nature makes both nature and second nature appear in its image. But it is not quite a standing reserve. For instance, a piece of land is a standing-reserve. When we visited the Escalante National Monument, a park ranger explained that a national park comes into existence through an act of Congress; a national monument comes into existence through an executive order of the president. The Obama administration made extensive use of the latter capacity. As I write, the Trump Administration is contemplating an unprecedented reversal of some of those orders. Either way, *this* terrain, this nature, this federally managed land, is a standing-reserve; either for the people or for the private developers favored by the current administration.

Third nature abstracts from the this-ness of a standing-reserve. It puts all standing-reserves on the same plane. They can be named, tagged, archived using standard protocols. They can then be compared, evaluated, promoted, annotated—also using standard protocols. For us, it was a simple matter of comparing campgrounds. But the same third nature can also become a terrain that turns all of nature and second nature into targets for financial investment. They can be acquired, traded, held in portfolios, borrowed against, and in various ways, hedged.

The rather astonishing thing about third nature is that any information on it anywhere is theoretically as proximate to any other piece of information, as to any still other piece of information. It breaks the habits of proximity of second nature. My son is thankful for this, as he can keep watching his Netflix shows from various campgrounds, so long as they have Wi-Fi, and these days they often do. Just as the convenience of a drill on hand masks the vast infrastructure that made such a scenario possible, so too third nature appears as if all the information of the world is readily available for us. This too masks the vast infrastructure of the extensive and intensive vector.

The Grand Canyon seems like the appropriate stop on our RV tour to conclude this essay, which has turned into something like that classic genre of American homework assignments for the fall school term: what I did on my summer vacation. The vast vertical cuts of the canyons reveal strata upon strata of geological time, stretching back to a time before settler civilization, indigenous civilization, before the human, even back before the time of biological life itself, which is a mere subset of geological time.

The Grand Canyon offers sublime immersion in the totality of geological time, a perception of nature far beyond the phenomenological perception and comprehension of the human. And yet that sublime experience of the inhuman appears to us mere humans as something mediated by an inhuman world that sometimes slips out of view. That inhuman world is that of second nature—and now, third nature—a world made of labor and technology, and made in two distinct strata. There was and there remains a second-nature stratum of Fordist production, of matter bent toward human convenience through the application of vast amounts of energy. And there is a third-nature stratum, which functions in a somewhat different manner.

Third nature may have arisen to resolve problems in the organization of second nature, but it is now the stratum that organizes all the others. As vertical and deep and slow as geological time and space might be, third nature is horizontal and thin and fast, wrapping the world in vectors that move information, and move the world toward that information. It offers a sublime sensation of quite a different order to the Grand Canyon. Even if it is now commonplace, there is still something awe-inspiring about snapping a picture of the Grand Canyon, posting it to Instagram from the lip of the rim, and observing one's friends comment on it.

While third nature appears to have made all of nature and second nature available for our convenience, perhaps it is rather the reverse. Maybe what it really does is arrange everything for *its* convenience, including us. While it gives us trinkets of recognition and morsels of information, it takes the mountain lion's share. It generates an information surplus out of even our leisure activities, here on the road in an RV, for the greater glory of Google and Apple and Amazon and the rest. There's now a stratum over and above the old Fordist, capitalist economy of making things and generating a profit out of exploiting labor that accumulates surplus information out of both labor and nonlabor.

The information vector of third nature binds space in unprecedented ways, creating an almost invisible stratum via which information organizes the world in its image, extracting a surplus enjoyed by rather novel forms of corporate power that may no longer be capitalist in the strict sense of the world. But while it binds space as never before, it has no demonstrated ability to bind time. There's no way of knowing how long the era of third nature will last, but compared to the geological time one glimpses in a moment of vertigo peering down into the canyon, it won't be long.

Composite image of Earth at night showing Europe, North Africa, and the Middle East, 2012.

[I]n 2012 . . . (the) Blue Marble Next Generation 2012, assembled from data collected by the Visible/Infrared Imager Radiometer Suite (VIIRS) on the Suomi NPP satellite in six orbits over eight hours. . . . They are composites of massive quantities of remotely sensed data collected by satellite-borne sensors. . . . This is not the integrating vision of a particular person standing in a particular place or even floating in space. It is an image of something no human could see with his or her own eye, not only because it's cloudless, but because it's a full 360-degree composite, made of data collected and assembled over time, wrapped around a wireframe sphere to produce a view of the Earth at a resolution at least half a kilometer per pixel—and any continent can be chosen to be in the center of the image. — Laura Kurgan, 2013

# Confronting the Popular Anthropocene: Toward an Ecology of Hope

Jason W. Moore

It's been shorts-and-T-shirt weather on Antarctica lately. Well, almost. This past spring, the World Meteorological Organization announced that temperatures on the polar continent reached 63.5°F on March 24, 2015. Barely two months later, we learned that Antarctica's massive Larsen C ice shelf was about to calve after a 110-mile crack advanced another 11 miles—in just six days. By the time this essay is published, the Southern Ocean will be home to a new iceberg the size of Delaware: some 2,000 square miles, 600 feet thick, weighing a trillion tons.[1]

Stories of spectacular biospheric change abound in the early 21st century. Ours is an era of epochal planetary transition—dramatic, irreversible, and chaotic. Earth scientists call such eras "state shifts."[2] That's a dry term for a reasonably terrifying situation. It means that the conditions of life will be fundamentally different within a generation. Sea levels will rise faster than anyone expected; stretches of the Middle East will become uninhabitable; and agriculture will be riskier as it is unmoored from its five-century model of eviscerating the earth for Cheap Food.[3] It means that my seven-year-old son will live his adult life—if he is fortunate—in an increasingly unstable and comparatively inhospitable biosphere.

If ever there was a pressing need for revolution, this is it. Capitalism—a dynamic crystallization of capital, power, and nature that has endured for five centuries—is now generalizing what it's always been for some: an intergenerational system of mass murder.

What stories do we need to make sense of this disastrous state of affairs, and to forge a politics of climate justice? One powerful answer is given by the Anthropocene, the most influential environmentalist concept of the past decade. Here I want to set aside the focused geological discussion, in which geologists debate stratigraphic markers. I've called this the Geological Anthropocene. The markers now settled upon are telling enough.[4] They amount to a searing indictment of postwar capitalism. Its signal accomplishments will be reduced in the geological record to a few millimeters of dust, identified by plastics, radioactive isotopes, and chicken bones. Lots of chicken bones.[5]

But it's not stratigraphy that's generated all the buzz; it's something we can call the Popular Anthropocene. It's *this* Anthropocene that asks—and answers—the most relevant questions of our times: What are the causes of today's planetary crisis? And when do we locate its origins?

These are the right questions. But the Popular Anthropocene limits our thinking about possible answers—and the stories they're embedded in—before we can really get started. By asking us to return to view of environmental problems premised on "humans" against "nature," this Anthropocene returns us to the thinking that created these crises in the first place. Far from an innocent binary, the binary of "man" and "nature" has been fundamental to colonial rule, environmental change, and genocide ever since Columbus landed on Hispaniola. The idea of Humanity as the agent of environmental crisis—today crystallized in the language of anthropogenic change—has been an indispensable weapon in capitalism's arsenal.

Anthropogenic (*made by humans*). Here we see an old capitalist trick playing out through environmentalist discourse: take a problem created by the 1 percent, then tell the 99 percent it's their fault. To credit humanity as the cause of climate change is to engage in a special brand of magical thinking. It says, in effect, that the inequalities and violence of race and class and gender are secondary concerns. Nor is it a speculative claim. Environmentalism emerged after 1968 as a politics of nature in which race, class, and gender were indeed secondary—when they were acknowledged at all. Nowhere is this expressed more clearly in Anthropocene discourse than in its deliciously neoliberal phrase "the human enterprise"—a term borrowed from Paul Ehrlich, whose 1968 *Population Bomb* drips with racial anxiety.

The principal driver of modern environmental change is not anthropogenic, but capitalogenic (*made by capital*). But that's not so simple as it first appears. Global warming is not propelled by "the economy" as such, although that's clearly implicated. Rather, to say *capitalogenic* is to indict a system of power, capital, and nature. That system weaves together a peculiar rationality—in science, in power, in economics—that compels endless expansion as an existential condition. In a world teeming with life that is not only finite—but unruly—such a logic is fraught with danger as well as possibility.

The Popular Anthropocene ignores all this. The structure of its thinking—Man versus Nature—compels such ignorance. An alternative is to recognize that the planetary "state shift" identified by earth system scientists requires an *intellectual* state shift. By privileging the "human enterprise," we return to habits of thought forged in the long 18th

---

1   Chris Mooney, "A 2,000 Square Mile Iceberg Is About to Break off of Antarctica," *The Independent*, July 8, 2017, http://www.independent.co.uk/news/science/iceberg-delaware-break-off-antarctica-global-warming-a7830416.htmlm. See especially the Antarctic research blog focused on the Larsen C shelf, "Project Midas: Impact of Melt on Ice Shelf Dynamics and Stability," http://www.projectmidas.org.

2   A. D. Barnosky, Elizabeth A. Hadly, Jordi Bascompte, Eric L. Berlow, James H. Brown, Mikael Fortelius, Wayne M. Getz, John Harte, Alan Hastings, Pablo A. Marquet, Neo D. Martinez, Arne Mooers, Peter Roopnarine, Geerat Vermeij, John W. Williams, Rosemary Gillespie, Justin Kitzes, Charles Marshall, Nicholas Matzke, David P. Mindell, Eloy Revilla, and Adam B. Smith, "Approaching a State Shift in Earth's Biosphere," *Nature* 486 (June 2012): 52–58.

3   See James Hansen, Makiko Sato, Paul Hearty, Reto Ruedy, Maxwell Kelley, Valerie Masson-Delmotte, Gary Russell, George Tselioudis, Junji Cao, Eric Rignot, Isabella Velicogna, Blair Tormey, Bailey Donovan, Evgeniya Kandiano, Karina von Schuckmann, Pushker Kharecha, Allegra N. Legrande, Michael Bauer, and Kwok-Wai Lo, "Ice Melt, Sea Level Rise and Superstorms: Evidence from Paleoclimate Data, Climate Modeling, and Modern Observations That 2°C Global Warming Could Be Dangerous," *Atmospheric Chemistry and Physics* 16, no. 6 (March 2016): 3761–812; Jeremy S. Pal and Elfatih A. B. Eltahir, "Future Temperature in Southwest Asia Projected to Exceed a Threshold for Human Adaptability," *Nature Climate Change* 6, no. 2 (February 2016): 197–200; Jason W. Moore, *Capitalism in the Web of Life: Ecology and the Accumulation of Capital* (London: Verso, 2015), 241–90.

4   Damian Carrington, "The Anthropocene Epoch: Scientists Declare Dawn of Human-Influenced Age," *The Guardian*, August 29, 2016, https://www.theguardian.com/environment/2016/aug/29/declare-anthropocene-epoch-experts-urge-geological-congress-human-impact-earth.

5   Jason W. Moore, "The Capitalocene, Part I: On the Origins of Our Ecological Crisis," *Journal of Peasant Studies* 44, no. 3 (2017): 594–630; idem, "The Capitalocene, Part II: Accumulation by Appropriation and the Centrality of Unpaid Work/Energy," *Journal of Peasant Studies* (online in advance of print, 2017).

century.[6] This isn't because of fossil fuels, or fossil capital—even if that remains important. Rather, it's because in this era we see the rise of the liberal subject, of "man" as the agent of improvement, ideologically cleansed of racialized, gendered, and colonial violence. The symbolic and the material are so entwined in the liberal subject that it's hard to see where one ends and the other begins. Such Lockean dualism marked Nature (in the uppercase) as the terrain of Society's improvement; untouched except by native peoples incapable of improvement. Here we find a crystallization of power, ideas, and nature no less significant than Manchester's textile factories.[7] There was no fusion of coal, steam, and cotton without exterminated and removed indigenous peoples, on whose unceded lands enslaved Africans worked, and worked to death. In privileging the human enterprise, the Popular Anthropocene marks the highest stage of dualist thought, and a return to habits of thought that served British imperialists well. The Popular Anthropocene, from this perspective, is peak liberalism.

The responsibility of the radical is to name the system. Naming is fundamental to any political project, and the power to name is the power to channel thought—and to shape emancipatory vistas. That the Anthropocene, at its core, is a fundamentally bourgeois concept should surprise no one. After all, it tells us that behind the current, disastrous state of world affairs is the *Anthropos*. The Popular Anthropocene is but the latest in a long series of environmental concepts that deny the multispecies violence and inequality of capitalism and assert that the devastation created by capital is the responsibility of all humans. The politics of the Anthropocene is in fact an anti-politics, resolutely committed to the erasure of capitalism and the capitalogenesis of planetary crisis.[8]

In this light, the Anthropocene may not be merely peak liberalism, but peak *neo*liberalism.

## Dualism and the Cartesian Revolution, Or, Capitalism as Mode of Thought

From the standpoint of the *longue durée*, the Popular Anthropocene is also peak dualism. It pushes Green Arithmetic—Nature plus Society —to the breaking point. I've often called that dualism Cartesian, after René Descartes's famous distinction between "thinking things" (brains, societies) and "extended things."

Of course Descartes's thinking was more sophisticated than the crude dualism suggests. But the issue is not exegetical; it is, rather, one of the uptake of intellectual systems into the world-historical apparatus of domination. The Cartesian system coevolved with emergent structures of capital and power to install a binary code at the heart of modernity. The centuries after 1492 witnessed not only the birth of a mode of production, but also a mode of thought. Long before the algorithmic revolutions of neoliberal surveillance, the capitalist mode of thought churned out one binary code after another, installing each in the systems of domination and exploitation characteristic of early capitalism: White and Black; Man and Woman; Colonizer and Colonized; Society and Nature. It's no coincidence that our familiar words Nature, Society, and European all assumed their present-day meanings in the century or so after 1550; an era of brutal colonization in Ireland and the Americas, murderous witch hunts and the violent regulation of female bodies, the first great boom of the African slave trade, the neocolonial subordination of Poland, interminable warfare, the continental-scale reorganization of Andean life and work, rapid deforestation from Brazil to the Baltic, and the vigorous spread of sugar plantations across the Western Hemisphere.

It's no coincidence that Descartes's formulation appeared toward the end of this period. If a cosmology of thinking and extending things was Cartesian, its inspiration was certainly Columbian and Cortesian. Capitalism as mode of thought was a mode of conquest. The praxis of Cheap Nature, a spiritual, bodily, and cultural degradation no less than a biophysical one, appears in this light as a putrid marriage of conquest and cognition. In this praxis, some humans (rich, white, male) were "thinking things." This was the realm of Society. Nature, on the other hand, was full of extended things: women, people of color, Indigenous peoples.

On this point, we can no longer feign ignorance: Descartes's philosophical abstractions were practical instruments of domination. Cartesian rationality became a theoretical hammer in the hands of empire. Nature and Society (or "Man") became *real abstractions* with tremendous material force.[9]

These are just a few reasons why the Anthropocene frame is an intellectual prison house. Modes of thought are tenacious. They are no easier to transcend than the "modes of production" they reflect and shape. My argument for the Capitalocene is precisely to encourage conversation—and the practical political struggle that implies—around a new mode of thought.

---

[6] The phrase "human enterprise" recurs throughout Will Steffen's writing. See Will Steffen, Paul J. Crutzen, and John R. McNeill, "The Anthropocene: Are Humans Now Overwhelming the Great Forces of Nature," *AMBIO: A Journal of the Human Environment* 36, no. 8 (2007): 614–21. For a searing critique, see Eileen Crist, "On the Poverty of Our Nomenclature," in *Anthropocene or Capitalocene?*, ed. Jason W. Moore (Oakland: PM Press, 2016), 14–33.

[7] Zahir Kolia, "The Capitalocene, Coloniality, and Theology: John Locke and British Colonialism in the World-Ecology," in *Capitalism's Ecologies*, eds. Jason W. Moore, Sharae Deckard, Michael Niblett, and Diana C. Gildea (Oakland: PM Press, 2017).

[8] On anti-politics, see James Ferguson, *The Anti-Politics Machine: "Development," Depoliticization, and Bureaucratic Power in Lesotho* (Cambridge: Cambridge University Press, 1990). For anti-politics in the Anthropocene, see Erik Swyngedouw, "Depoliticized Environments: The End of Nature, Climate Change, and the Post-Political Condition," *Royal Institute of Philosophy Supplements* 69 (2011): 253–74.

[9] For a brilliant reflection on real abstraction, see Alberto Toscano, "The World Is Already Without Us," *Social Text* 34, no. 2 (2016): 109–24.

# The Geological Anthropocene and the Capitalocene as Geohistorical Unity

The Popular Anthropocene is an intellectual prison house—one whose intellectual lineage reaches into capitalism's deep history of *actual* imprisonment, enslavement, and enclosure.

If not the Anthropocene, what? In a word, the Capitalocene. That doesn't mean a return to the primacy of economic motives. If we are to grasp the nettle of the biospheric state shift, we'll need something more nuanced than economic determinism and Green Arithmetic allow. We'll need a revolutionary mode of thought capable of informing a politics of justice—for humans, for the biosphere, and for future generations.

As I often say, Capitalocene is an ugly word, an ugly system. Then again, "Anthropocene" isn't exactly Shakespeare. In its concern for planetary crisis, the Capitalocene foregrounds a certain kinship with the Popular Anthropocene. But the Capitalocene refuses the mode of thought that tells us the world can be understood, and acted upon, through the doubly violent abstractions of Humanity and Nature. The bloodstains on these concepts—and the system of Cheap Nature that produced them—simply cannot be bleached out.

The elements of an alternative mode of thought are now emerging and will require an ongoing movement of intellectual disobedience to authoritarian structures of disciplined knowledge. Eco-socialists have so far affirmed their allegiance to Green Arithmetic. Political ecologists have so far refused the world-historical. Post-sructuralists have kept their distance from the political economy of capital accumulation. If the generals are always prepared to fight the last war, radicals are always prepared to organize the previous revolution. This will not suffice in an era of planetary state shifts.

I and others in the world-ecology conversation have argued for a dialectical perspective on humans in the web of life—humans make natures, and the web of life makes us. "Us" in this instance is an ontologically differentiated unity, to be sure. Dialectics is one of the least understood ideas in critical theory. A minimally dialectical program unfolds from two observations. One is the intimate, porous, and interpenetrating relations of the "parts" in the whole. For eco-socialists, this means recognizing how the web of life produces the circuit of capital, even as capital transforms the biosphere. A second is that the parts acquire new properties through the historical process. That's an abstract way of saying something simple: capitalism is not structurally invariant but dynamic as it develops through the web of life; the web of life is equally dynamic (more so!) and experiences significant changes as it moves through capitalist relations of power and re/production. Like all dialectical processes, that's an asymmetrical relationship whose terms of dominance and subordination shift over time.[10]

It's a relationship characterized by a "double internality": the biosphere internalizes the contradictions of capitalism as capitalism internalizes the twists and turns of planetary history.[11] The Geological Anthropocene produces (and is produced through) the Capitalocene, though never equally through time and space.[12] The philosophical perspective powerfully shapes how one sees the knotty relations of capitalist and planetary crisis. What the Capitalocene refuses is the reduction of biospheric crisis to economic contradiction. It says that capitalism is best conceptualized as a world-ecology that joins the pursuit of power, the accumulation of capital, and the coproduction of nature as a historical unity. The relations of capitalism and the biosphere are fundamentally interpenetrated.

It is insufficient to conceptualize and theorize these relations; they must be historicized and situated geographically. The Capitalocene, like the Popular Anthropocene, insists on a deep historical perspective. It equally insists that we all become geographers—"earth writers." For the Capitalocene's deep history, from its Columbian origins, reveals the space of capitalist transition—and the origins of today's crisis—as a Pangea reunified by the era's militarized and racist commodity regimes.

Capitalism will no more survive the ongoing state shift than the Roman Empire or feudal Europe survived far milder climate changes in their respective eras. One of the problems of starting the world-historical clock in 1800 is that you forget a much longer history. If you start thinking of capitalism through the steam engine, you end up with a rosy picture of capitalism's resilience. Change the technology, change the energy source, save the planet. Shift the optic to capitalism as a system of Cheap Nature with a limitless appetite for work of all natures and a sociopathic disregard for their care, and the picture changes. The steam engine becomes but one crucial instrument of capitalist power, profit, and its capacity for latent and actual mass murder.

It is something of an article of radical faith these days that capitalism can survive the ongoing state shift. That seems unlikely. For starters, today's climate change, by any measure, overshadows by some margin the two most recent climate transitions: between Roman Warm Period and Dark Age Cold Period (300–500 CE), and between the Medieval Warm Period and the Little Ice Age (1250–1450). Both were relatively mild by today's standards, and both were implicated in civilizational crises. Today's climate shift is far greater than the end to an era of favorable climate, such as that enjoyed by the Romans or feudal lords in their respective golden ages. It's an end to the *whole era of favorable climate since the dawn of civilization*.

The Capitalocene argument agrees that significant upticks in $CO_2$ emissions occurred after 1850, and again after 1945. It differs from the Popular Anthropocene in arguing that the sources of planetary crisis are the relations of power and re/production that developed in the centuries after 1492. No doubt these relations are enormously complex. But we can capture something of their essence in what I've called the praxis of Cheap Nature.

---

10  Jason W. Moore, "Metabolic Rift or Metabolic Shift?: Dialectics, Nature, and the World-Historical Method," *Theory & Society* (forthcoming 2017).

11  Moore, *Capitalism in the Web of Life*, 1–30.

12  This is the implicit argument made by Mark A. Maslin and Simon L. Lewis, "Anthropocene: Earth System, Geological, Philosophical and Political Paradigm Shifts," *Anthropocene Review* 2, no. 2 (2015): 108–16.

Cheap Nature works through a basic principle reduce all work—including the work of nature as a whole—to its most simplified forms and its most basic qualities. Capitalist agriculture has been dominated by monocultures for this reason. It's this principle that's given us assembly lines with radically simplified work motions, and "flex crops" like maize that can be readily converted into food, fuel, or raw material. This principle shapes two logics, each essential to the other. The first turns on cheapening in a basically Marxist sense, reducing production costs to the bare minimum. In this view, great booms of capitalist development have turned on the extra-economic appropriation of unpaid human and extra-human work. In this sense, frontiers of uncapitalized nature—including human nature—are indispensable to capitalist survival. Our second logic hinges on domination and cheapen*ing*: reducing the work and lives of women, people of color, and indigenous people to the lowest possible cultural priority. The epochal redefinition of "women's work" as "nonwork" in the early modern centuries stands as a signal moment of such cheapening.

Capitalism as a system of Cheap Nature remade life, land, and sea long before the Industrial Revolution. Indeed, the centuries after Columbus landed on Hispaniola marked an epochal rupture in human-initiated environment-making, unprecedented since the dawn of agriculture and the rise of the first cities. The massive infrastructures of empire and capital that soon emerged, marking the first great wave of planetary urbanization, effectively reunified Pangea for the first time in 180 million years.[13] Suddenly, the work/energy potential of two continents could be appropriated for Europe's capitalist empires. Across the early modern period fields were planted, forests cleared, indigenous peoples exterminated, mines dug, metals smelted, peasants dispossessed—all at scale, scope, and speed that exceeded, often by an order of magnitude, the standards of premodern civilizations.

What the Capitalocene foregrounds is the intimate and dialectical connection between work and nature. The early modern landscape revolution was also a revolution in labor productivity. Indeed, the violence of Cartesian dualism has prevented the elementary synthesis offered by recognizing that advancing labor productivity is, centrally, the increasing material throughput of natures for every hour of labor-time. There is no Global Factory without a Global Farm and Global Mine—and none exist without the Global Household of unpaid care.[14] This complex set of world-historical relations should remind us that capitalist revolutions always join commodification with revolutions in extra-economic appropriation necessary to reestablish the Four Cheaps of food, labor, energy, and raw materials.[15]

Imperialism has been central to life and times of the Four Cheaps. But there's more to it than brute force and vulgar materialism. Empires rule through cultures and epistemologies no less than guns and dollars. Early modernity's "new" imperialism was fundamentally enabled by the Cartesian revolution, which established a new way of seeing and ordering planetary life. (Has it been so different for the new imperialism of the neoliberal era?) One could conquer the globe only if one could imagine it. Early forms of external nature, abstract space, and linear time enabled capitalists and empires to construct global webs of exploitation and appropriation, calculation and credit, property and profit, on an unprecedented scale.

These in turn were fundamental to capitalism's real basis of profit: labor productivity. While many environmentalists and Marxists continue to insist that real capitalism begins only in the nineteenth century, they've missed out on early capitalism's signal accomplishment: its crystallization of force, commerce, and rationality to transform all of planetary life into a potential condition of capital accumulation. In one key sector after another—shipping and shipbuilding, sugar planting, silver and copper mining, textiles, cereal agriculture—labor productivity surged, and surplus value flooded onto the ledgers of merchants, bankers, and industrialists. The problem—then as now—was not too little capital, but too much.

Early capitalism's fundamental problem—in contrast to the political economy of the long fossil boom that has shaped the past two centuries—was not too few customers, but too few inputs. The danger was not *over*production, but *under*production. "Aha!" says the environmentalist. Are we not back to Malthus and the limits to growth? Yes and no—or rather, no and yes. No to Malthus. But yes to Marx's relational concept of underproduction, in which capital's industrial dynamism tends to outstrip capitalism's capacity to supply Cheap energy and raw materials. The greater the industrial dynamism, the greater the possibility that input costs will rise, the greater the profit squeeze. This was precisely the situation in which the early modern accumulators of capital found themselves. This turns inside-out the usual discussion of early capitalism: Not why and how did early capitalism industrialize so little, but why and how did it industrialize so much?

The answer turns on a historically unusual combination of money, trade, and power that took shape in the century after 1450. The three great conquests of the late 15th century—the final subordination of the Canary Islands, the defeat of Granada, and the Columbian invasion—were largely financed by Genoese capitalists. Like any good capitalist, the Genoese were interested in one thing: profit. But the sources of immediate profit were not obvious, except one: slaving. Empires needed cash; capitalists needed profits; planters needed workers. Slaving, and the wider structures of coercive labor control, addressed all these needs. The die was cast.

It is in this sense that something seemingly as innocuous as labor productivity become a world-historical affair. Early capitalism's labor productivity revolution turned, in short, on a Great Frontier opened and sustained through genocide and conquest. Thus worldwide landscape transformation was fundamental to a labor productivity revoltuion. These were frontiers repopulated through the violence of the Great Domestication: the violent subordination of women in Old and

---

13  On planetary urbanization, see Neil Brenner, *Critique of Urbanization* (Gütersloh: Bauverlag, 2017). On the reunification of Pangea, see Alfred W. Crosby, *Ecological Imperialism: The Biological Expansion of Europe, 900–1900* (Cambridge: Cambridge University Press, 1985).

14  The global household concept originates with Maliha Safri and Julie Graham, "The Global Household: Toward a Feminist and Postcapitalist Political Economy," *Signs* 36, no. 1 (2010): 99–125. On the centrality of care work to historical capitalism, see Patel and Moore, *Seven Cheap Things*.

15  Moore, *Capitalism in the Web of Life*, 91–165.

New Worlds. The fact that early capitalism relied on global expansion—operationalized through racialized and gendered dualisms—to advance labor productivity reveals early capitalism's remarkable precocity, not its premodern character. This precocity allowed early capitalism to defy the premodern pattern of boom and bust. There would be no system-wide reversal of commodification after 1450, not during the "crisis" of the 17th century, and not during the great era of worldwide revolt and revolution after 1776. Why? Because early capitalism's *technics*—its crystallization of machines and power, knowledge and production—were specifically organized to treat the appropriation of global and household space as the basis for the accumulation of wealth in its modern form: capital as abstract social labor.

      The point that merits underlining is the degree to which we can draw rough and ready parallels between the early modern centuries and the 21st century. Like early capitalism, we appear to be entering an era when underproduction crises will decisively shape capital accumulation. Runaway climate change not only spells disaster for humans and the rest of life on this planet; it also signals a dramatic reversal of the fundamental enabling condition of capitalist develoment: the capacity to cheapen natures of all kinds and to advance labor productivity by absorbing new frontiers. Never mind the end of terrestrial and aquatic frontiers, mightily important in their own right. The enclosure of the atmosphere as a dumping ground for greenhouse gas emissions is *already* undermining agriculture's biological productivity, suppressing at once yield growth, nutritional content, and labor productivity. The early modern experience tells us that one tool above all is deployed in moments of underproduction crisis: force. And it's here that we find a profound disjuncture with the earlier history: in a world where capital has nowhere to run and precious few frontiers to enclose, and where indigenous, food, and climate justice movements have proliferated, capital's ability to rule by force is more compromised than ever.

## Ecologies of Hope

What stories do we need to find our way through the planetary state shift? It seems to me that such stories will need as touchstones the care, compassion, and connection that is so deeply lacking in today's world. The reflections I've offered, though framed as an argument for an intellectual state shift, must reach beyond the intellectual—another category violently reshaped by the Cartesian revolution. How do we act and think and love and organize our way through this state shift? Central to world-ecology has been the argument that we need to think, work and nature in new ways—especially through an ethic of care, for humans, for the web of life, and for the multispecies interdependencies that make the good life possible. That means putting nature at the center of thinking about work; putting work at the center of our thinking about nature; and setting aside the presumption that human organization of any kind (from family forms to transnational corporations) can be adequately understood abstracted from the web of life. From these intellectual—but always more than intellectual—ruptures, we may find ways toward to converse and cultivate and care for new ecologies of hope and justice in the 21st century.

Oceanic and atmospheric dynamism, 2002.

We provide for ourselves transcendental figurations of what we think is the origin of this animating gift: mother, nation, god, nature. These are names of alterity, some more radical than others. Planet-thought opens up to embrace an inexhaustible taxonomy of such names, including but not identical with the whole range of human universals: aboriginal animism as well as the spectral white mythology of postrational science. If we imagine ourselves as planetary subjects rather than global agents, planetary creatures rather than global entities, alterity remains underived from us; it is not our dialectical negation, it contains us as much as it flings us away. And thus to think of it is already to transgress, for, in spite of our forays into what we metaphorize, differently, as outer and inner space, what is above and beyond our own reach is not continuous with us as it is not, indeed, specifically discontinuous. — Gayatri Spivak, 2003

## Visual Essay Citations and Image Credits

The visual narrative that appears in the space in-between essays is an incomplete attempt by the editors to represent—via the articulation of variegated visual and textual sources—the wide spectrum of geospatial configurations through which the contemporary posthuman condition described in the volume is materialized, imagined, and contested at multiple scales.

**18–19**
**Citation**
Nicholas de Monchaux, *Spacesuit: Fashioning Apollo* (Cambridge, MA: MIT Press, 2011), 337.
**Image Credit**
Photo by NASA on Unsplash

**26–27**
**Citation**
David Harvey, *Abstract from the Concrete* (Cambridge, MA: Stenberg Press, Harvard GSD, 2016), 9.
**Image Credit**
Aurélien Maréchal

**32–33**
**Citation**
N. Katherine Hayles, *How We Became Posthuman: Virtual Bodies in Cybernetics, Literature, and Informatics* (Chicago: University of Chicago Press, 1999), 290.
**Image Credit**
Reuben Wu

**42–43**
**Citation**
Nick Bostrom, *Superintelligence: Paths, Dangers, Strategies* (Oxford: Oxford University Press, 2014), 4–5.
**Image Credit**
Michael Hansmeyer

**62–63**
**Citation**
Rosalind Williams, *Notes on the Underground: An Essay on Technology, Society, and the Imagination* (Cambridge, MA: MIT Press, 2008), 212–13.
**Image Credit**
Photo by Vincent Erhart on Unsplash

**68–69**
**Citation**
Adam Greenfield, *Radical Technologies: The Design of Everyday Life* (London and New York: Verso Books), 28.
**Image Credit**
Photo by Redd Angelo on Unsplash

**78–79**
**Citation**
Pierre Bélanger, *Extraction*, Exhibition Catalog, Venice Architecture Biennale, Canadian Exhibition, 2016.
**Image Credit**
Jose Ahedo

**96–97**
**Citation**
Achille Mbembe, "Necropolitics," *Public Culture* 15, no. 1 (2003): 34.
**Image Credit**
Photo by SpaceX on Unsplash

**106–107**
**Citation**
Neil Brenner, *Implosions/Explosions: Towards a Study of Planetary Urbanization* (Berlin: Jovis, 2014), 26.
**Image Credit**
Photo by Mariusz Prusaczyk on Unsplash

**114–115**
**Citation**
Saskia Sassen, *Expulsions: Brutality and Complexity in the Global Economy* (Cambridge, MA: The Belknap Press of Harvard University Press, 2014), 2.
**Image Credit**
Benjamin Grant, Daily Overview

**138–139**
**Citation**
Ian Bogost, *Alien Phenomenology, or What It's Like to Be a Thing* (University Press of Minnesota, 2012), 73.
**Image Credit**
Photo by Mariusz Prusaczyk on Unsplash

**148–149**
**Citation**
Leanne Betasamosake Simpson, "Land as Pedagogy: Nishnaabeg Intelligence and Rebellious Transformation," *Decolonization: Indegeneity, Education & Society* 3, no. 3 (2014): 22.
**Image Credit**
Chief Lady Bird (Nancy King), with Aura (Monique Bedard) and the grade 5's and 6's students at First Nations School

**154–155**
**Citation**
Zygmunt Bauman, *Globalization and Human Consequences* (Cambridge: Polity Press, 1998), 2.
**Image Credit**
Cassio Vasconcellos

**162–163**
**Citation**
Donna Haraway, "A Cyborg Manifesto," *Berkeley Socialist Review Collective* (1985): 295.
**Image Credit**
Benjamin Grant, Daily Overview

**174–175**
**Citation**
Timothy Morton, *Hyperobjects: Philosophy and Ecology after the End of the World* (Minneapolis: University of Minnesota Press, 2013), Kindle Edition, Part II: The Time of Hyperobjects, Chapter: The End of the World.
**Image Credit**
Reuben Wu

**192–193**
**Citation**
Laura Kurgan, *Close Up at a Distance: Mapping, Technology, and Politics* (Brooklyn: Zone Books, 2013), 11.
**Image Credit**
Photo by NASA Earth Observatory

**200–201**
**Citation**
Gayatri Chakravorty Spivak, *Death of Discipline* (New York: Columbia University Press, 2003), Kindle Edition, Chapter 3: Planetary.
**Image Credit**
Photo by NASA Goddard Space Flight Center, image by Reto Stöckli, enhancements by Robert Simmon

## Biographies

**Barbara Adam** is Emerita Professor at Cardiff University's School of Social Sciences. The social temporal has been her intellectual project throughout her academic career, resulting in five research monographs and a large number of articles in which she sought to bring time to the center of social science analysis. This focus facilitated a unique understanding, whose relevance transcends perspectives and disciplines. On the basis of this work she has been awarded two book prizes and several social theory-based research Fellowships and grants. She is founding editor of the journal *Time & Society*.

**Jose Ahedo** founded Studio Ahedo in 2010, specializing in residential design and consultation for graphic design, branding, and software development. The first completed Studio Ahedo project is Blanca, a dairy complex of 13 buildings in the Pyrenees mountains including animal facilities, research labs, and an education center. Ahedo's proposal "Domesticated Grounds: Design and Domesticity within Animal Farming Systems" was awarded the Harvard University Graduate School of Design 2014 Wheelwright Prize.

**Pedro Aparicio Llorente** is a Colombian architect and cofounding partner of both Altiplano and CAMPO; the latter is an architecture office and research studio based in Bogotá that overlaps construction techniques with geographic inquiries. He has been a faculty member at the Rhode Island School of Design, visiting researcher at the Rujak Center for Urban Studies in Jakarta, Indonesia, and research associate at the Office for Urbanization at Harvard University Graduate School of Design. Aparicio investigates socio-spatial patterns that emerge within tropical regions. He holds an MDes from the Harvard GSD that was supported by a Fulbright fellowship.

**Martín Arboleda** is a political geographer based at the Urban Theory Lab-Graduate School of Design, Harvard University. His research engages with critical political economy, urban political ecology, and science and technology studies. He is currently working on a book project on resource extraction in the context of the new international division of labor. His work has been published by *Geoforum*, *Antipode*, *Environment and Planning D: Society and Space*, and *International Journal of Urban and Regional Research*, among other journals.

**Benjamin H. Bratton**'s work spans philosophy, art, design, and computation science. He is Professor of Visual Arts and Director of the Center for Design and Geopolitics at the University of California, San Diego. He is Program Director of the Strelka Institute for Media, Architecture, and Design in Moscow, Professor of Digital Design at the European Graduate School, and Visiting Faculty at Southern California Institute of Architecture (SCI-Arc). Recent publications include *The Stack: On Software and Sovereignty* (2016) and *Dispute Plan to Prevent Future Luxury Constitution* (2015).

**John Dean Davis** is an environmental and design historian, and a PhD candidate at Harvard University. His dissertation is a history of landscape and state engineering in the American South during the Reconstruction of the American Landscape (1865–1885).

**Krystelle Denis** is a designer, technologist, and researcher who creates visual frameworks and data-driven narratives to both represent and organize material and digital abundance. She uses both design and coding to produce a variety of digital artifacts, experimenting with visualizations, maps, and interactive documentaries. She holds an MArch from Harvard University Graduate School of Design, and currently works as a design researcher at metaLAB at the Harvard and Labex OBVIL at Paris-Sorbonne University.

**Rosetta S. Elkin** is Assistant Professor of Landscape Architecture at the Harvard University Graduate School of Design, research associate at Harvard's Arnold Arboretum, and co-director of the Master in Design Studies Risk and Resilience concentration. Elkin's current research explores the geopolitical practices of supracontinental tree-planting programs. Her work has been featured internationally, including projects at Les Jardins de Metis, Chelsea Festival, and the Isabella Stewart Gardner Museum. Her recent publication *Tiny Taxonomy* (2017) deliberates on scales of practice by exploring the role of individual species.

**GIDEST** is a research and exploration lab at the New School in New York. More information on the 2016–2017 collective who co-authored "Living Past the End Times" can be found at http://www.gidest.org/2017-fellows.

**Mariano Gomez-Luque** is a practicing architect and urban designer from Argentina, a Doctor of Design candidate at the Harvard University Graduate School of Design, and a research fellow at both the Urban Theory Lab and the Office for Urbanization. His work investigates the complex relations between architecture, planetary urbanization, and associated processes of technological change; and the production of space at multiple scales under contemporary capitalism. He holds an MArch with distinction from Harvard GSD.

**Stephen Graham** is Professor of Cities and Society in the Global Urban Research Unit based at Newcastle University's School of Architecture, Planning, and Landscape. He has been Visiting Professor at the Massachusetts Institute of Technology (MIT) and New York University (NYU), among other institutions. He is author, editor, and coauthor of seven books including *Vertical: The City from Satellite to Bunkers* (2016), *Cities under Siege* (2010), and *Splintering Urbanism* (2001).

**Ghazal Jafari** is an architectural designer and researcher. She is currently a Doctor of Design candidate at the Harvard University Graduate School of Design and a research fellow at the Weatherhead Center for International Affairs. Her work is situated at the intersection of the ethnography of infrastructure, and geographic formations, territoriality, and landscapes of neocolonialism. Her dissertation examines the geopolitical influence of global corporations and market-driven territories of (im)mobility. She holds an MDěs from Harvard GSD.

**Namik Mackic** is a design researcher and strategic designer who investigates collective, landscape-specific alternatives to colonial and modernist urban systems and forms, with a broad interest in mobility across geopolitical, demographic, and environmental dynamics. He holds an MDes from the Harvard University Graduate School of Design, where he was the inaugural recipient of the 2017 Richard Rogers Fellowship. He has been a guest lecturer at Rhode Island School of Design.

**Shannon Mattern** is Associate Professor of Media Studies at the New School. Her writing and teaching focus on archives, libraries, and other media spaces; media infrastructures; spatial epistemologies; and mediated sensation and exhibition. She is the author of *Code and Clay, Data and Dirt: Five Thousand Years of Urban Media* (2017); *Deep Mapping the Media City* (2015); and *The New Downtown Library: Designing with Communities* (2007). She contributes a regular column on urban data and mediated infrastructures to *Places Journal* and also publishes at wordsinspace.net.

**Jason W. Moore** is a world historian and geographer at Binghamton University. He is author of *Capitalism in the Web of Life: Ecology and the Accumulation of Capital* (2015); and, with Raj Patel, coauthor of *A History of the World in Seven Cheap Things* (2017). Moore coordinates the World-Ecology Research Network, https://worldecologynetwork.wordpress.com.

**Eli Nelson** is a PhD candidate in the History of Science at Harvard University, a doctoral fellow in the Science, Religion, and Culture program at Harvard Divinity School, and a graduate student associate of the Weatherhead Center for International Affairs. He studies the history of native science in settler-colonial and postcolonial contexts. His dissertation traces the history of 20th-century indigenous engagements with hegemonic Western science in the United States.

**Luciana Parisi** is Reader in Cultural Theory, Chair of the PhD Program at the Centre for Cultural Studies, and codirector of the Digital Culture Unit at Goldsmiths, University of London. She is the author of *Contagious Architecture: Computation, Aesthetics, and Space* (2013) and *Abstract Sex: Philosophy, Biotechnology, and the Mutations of Desire* (2004). Her current research examines the philosophical consequences of logical thinking in machines.

**Antoine Picon** is G. Ware Travelstead Professor of the History of Architecture and Technology and Director of Research at the Harvard University Graduate School of Design. He is the author of several books, including *Smart Cities: A Spatialized Intelligence* (2015), *Digital Culture in Architecture* (2010), and *French Architects and Engineers in the Age of Enlightenment* (1988; English translation, 1992).

**Carlo Ratti** is director of the Senseable City Laboratory at the Massachusetts Institute of Technology (MIT) and founding partner of the international design and innovation office Carlo Ratti Associati. He holds an MSc in Engineering from the Politecnico di Torino, Italy, and the Ecole des Ponts, France, and both an MPhil and PhD in Architecture from the University of Cambridge, UK. He attended MIT as a Fulbright scholar. His research interests include urban design, human-computer interfaces, electronic media, and the design of public spaces.

**Mimi Sheller** is Professor of Sociology and founding director of the Center for Mobilities Research & Policy at Drexel University, Philadelphia. She is President of the International Association for the History of Transport, Traffic, and Mobility, and founding coeditor of the journal *Mobilities*. Her books include *Aluminum Dreams: The Making of Light Modernity* (2014); *Mobility and Locative Media: Mobile Communication in Hybrid Spaces* (2014); *The Routledge Handbook of Mobilities* (2013); *Citizenship from Below: Erotic Agency and Caribbean Freedom* (2012).

**Erik A. M. Swyngedouw** is Professor of Geography at the University of Manchester. His research focuses on political ecology and political economy, with a particular interest in theorizing the society-nature relation through broadly historical-geographical materialist analysis. He has worked on urban socio-ecological dynamics, urban governance, politics of scale, and the geographical impact of advanced capitalist society. He is the author of *Liquid Power: Contested Hydro-Modernities in 20th-Century Spain* (2015) and *The Post-Political and Its Discontents: Spaces of Depoliticization, Spectres of Radical Politics* (2014). His new book, *Promises of the Political*, is forthcoming in 2018.

**McKenzie Wark**'s books include *Molecular Red: Theory for the Anthropocene* (2015); *The Beach Beneath the Street: The Everyday Life and Glorious Times of the Situationist International* (2011); and *Gamer Theory* (2007). He teaches at the New School in New York City.

**Cary Wolfe** is Bruce and Elizabeth Dunlevie Professor of English at Rice University, where he is also founding director of 3CT: Center for Critical and Cultural Theory. His recent books include *Before the Law: Humans and Other Animals in a Biopolitical Frame* (2013) and *What Is Posthumanism?* (2010). He is founding editor of the Posthumanities series published by the University of Minnesota Press.

**Charles Waldheim** is a North American architect, urbanist, and educator. Waldheim's research examines the relations between landscape, ecology, and contemporary urbanism. He coined the term "landscape urbanism" to describe the emergent discourse and practices of landscape in design culture and contemporary urbanization. He is the author of *Landscape as Urbanism: A General Theory* (2016) and editor of *The Landscape Urbanism Reader* (2006). Waldheim is John E. Irving Professor at the Harvard University Graduate School of Design and director of the Office for Urbanization.

**Eyal Weizman** is an architect, professor of spatial and visual cultures, and director of the Centre for Research Architecture at Goldsmiths, University of London. He was appointed global professor at the Princeton University School of Architecture in 2014. He established the research agency Forensic Architecture in 2010. Recent books include *Forensic Architecture: Violence on the Threshold of Detectability* (2017); *Hollow Land: Israel's Architecture of Occupation* (2007). He is on the editorial boards of *Third Text*, *Humanity*, *Cabinet*, and *Political Concepts*; and a board member of the Center for Investigative Journalism (CIJ), the Institute for Contemporary Art, London, and B'Tselem, Jerusalem, among others.

**Rosalind Williams** is a historian whose source for evidence and insight into the history of technology is imaginative literature. She is Bern Dibner Professor of the History of Science and Technology at the Massachusetts Institute of Technology (MIT), where she has taught since 1982. She has served as head of the Program in Science, Technology, and Society, and Dean of Undergraduate Education and Student Affairs at MIT; and President of the Society for the History of Technology. Williams is the author of several books, including *The Triumph of Human Empire: Verne, Morris, and Stevenson at the End of the World* (2013); *Retooling: A Historian Confronts Technological Change* (2002); and *Notes on the Underground: An Essay on Technology, Society, and the Imagination* (1990).

**Alejandro Zaera-Polo** is a practicing architect and founder of Alejandro Zaera-Polo & Maider Llaguno Architecture (AZPML), a practice based in London and New York. He is Professor of Architecture at the Princeton University School of Architecture and codirector of the 2017 Seoul Architectural Biennial. Zaera-Polo is coeditor of *Imminent Commons: Urban Questions for the Near Future* (2017) and *What Is Cosmopolitical Design? Design, Nature, and the Built Environment* (2016); and the author of the article "Well into the 21st Century. The Architectures of Post-Capitalism?" (2015), published in *El Croquis*.

**FORTHCOMING**
**New Geographies 10: Fallow**
Referencing the agricultural practice of rotational grazing known as fallowing, *New Geographies 10* proposes *fallow* as a metaphor to critically explore the dynamic states of devalorization and revalorization central to broader processes of contemporary capitalist urbanization. Through this process of fallowing, land is circulated into and out of use by capital as accumulation processes break down, while at the same time the vacuum of retreating capital provides fertile ground for the development of alternative modes of social production or appropriation by nonhuman actors.

**Editors**
Julia Smachylo & Michael Chieffalo